Real-Time Biomolecular Simulations

The Behavior of Biological Macromolecules from a Cellular Systems Perspective

Michael H. Peters, Ph.D.

McGraw-Hill

New York Chicago San Francisco
Lisbon London Madrid Mexico City
Milan New Delhi San Juan
Seoul Singapore Sydney Toronto

The McGraw·Hill Companies

McGraw-Hill books are available at special quantity discounts to use as premiums and sales promotions, or for use in corporate training programs. For more information, please write to the Director of Special Sales, Professional Publishing, McGraw-Hill, Two Penn Plaza, New York, NY 10121-2298. Or contact your local bookstore.

1234567890 DOC DOC 019876

ISBN-13: 978-0-07-146071-2
ISBN-10: 0-07-146071-3

Sponsoring Editor	Kenneth P. McCombs
Editorial Supervisor	Janet Walden
Project Manager	Vasundhara Sawhney
Acquisitions Coordinator	Laura Hahn
Copy Editor	Beatrice Sussman
Proofreader	Ojen Yumnam
Indexer	Robert Swanson
Production Supervisor	James Kussow
Composition	International Typesetting and Composition
Illustration	International Typesetting and Composition
Art Director, Cover	Handel Low

Contents

About the Author

Michael H. Peters, Ph.D., is a teacher/researcher in biomedical engineering and molecular biophysics. His particular focus is on the computational molecular biophysics of cell signal molecules (ligands) and their interactions with specific cellular targets, and clinical applications of ligands-receptor systems. The author of a textbook and over 30 journal articles, he is a frequent speaker at conferences and universities throughout the United States.

Preface

Living cells and tissues are highly dynamic entities that carry out complex interconnected processes and functions that depend on both space and time. Dynamic analysis is required to connect cellular structure to function and to predict cellular behavior under different "environmental" conditions and stresses. Furthermore, many important cellular processes, such as chemotaxis, feed-back control loops, and clock reactions, are well-known to be non-linear dynamic events that further complicate analytical descriptions.

It is clear that a quantitative analysis of cellular behavior that connects structure to function requires a systematic analysis of the particular unit processes and their interactions in the living cell. This book was written in order to lay out the basic systems engineering approach to quantitatively describe the dynamic cell. An introduction to the essential biochemistry in living systems is given in Chapter 1, followed by a description of the particular engineering systems of the living cell in Chapter 2. These systems are

- Cellular communications and transport
- Biochemical production
- Motility, mechanical work, and structural support
- Reproduction or division
- Information systems, checkpointing, and reliability

The majority of cellular processes are carried out by so-called macromolecules, such as peptides and proteins, lipids, carbohydrates, and nucleotides and oligonucleotides. Although this may appear to complicate the description of the living cell, in some ways it simplifies things. In short, the overall dynamical description of cellular processes involves large spatial and time scale separations between the behavior of the cellular solvent, i.e., water, dissolved ions, and small molecules, and that of the "workhorse" biological macromolecules. Chapters 3 and 4 develop the so-called multiple time scale methods that are used to exploit this disparity and simplify real-time macromolecular dynamics.

Finally, the last chapter of this book demonstrates the integration of system pathways to describe a cascade of cellular events that may lead to a particular phenotype. Although a full characterization of cellular processes along such rigorous lines is still a formidable goal, the last chapter explores some simplified examples of interconnected cellular system processes in order to illustrate the essential features of a rigorous approach in a revealing, analytical way.

In the future, large-scale computations will allow for the addition of more pathways and interconnectivity between systems in order to approach a more meaningful or useful total simulation of the dynamic living cell.

Acknowledgments

Thanks to Mr. Jack Schreiber at Virginia Commonwealth University for proofreading the final copy chapters of this book. Special thanks to my wife, Karen, and daughter, Emily, for letting me work a bit on weekends.

Chapter One

The World of Biological Macromolecules

The complexities associated with living systems are as far-reaching and as lacking understanding today as any field of science. Moreover, no single subject is as critical to our existence, including our interactions with our environment, as the dynamics and functionality of living systems.

Structure of Biological Molecules

For the most part, living systems can be broken down into the description of the molecules comprising them and, in particular, the dynamics, functionalities, and interactions of those molecules. Many molecules comprising living systems are actually large groups of smaller building-block molecules. For example, peptides and proteins represent large groups of smaller building-block molecules called *amino acids,* and starches and carbohydrates (the polysaccharides) represent groups of molecules called *simple sugars.*

Molecules representing large groups of smaller subunits often are referred to as macromolecules. Some macromolecules can be relatively large, such as proteins that can have molecular weights averaging over 30,000 as compared to, say, water with a molecular weight of only 18 Daltons.

Biological macromolecules represent the "molecular workhorses" of the living cell. They carry out communication as signaling molecules and produce other biochemical molecules as reaction catalysts, provide mechanical structure and motility, store information via their particular molecular arrangements, direct reproduction, and function as cellular "police" eager to arrest and dispose of aberrant molecules. Their world is our world, and the totality of their behavior and environmental interaction manifests into who we are.

Our story begins deceptively simple. In general, there are just four basic classifications of biological molecules:

- amino acids
- carbohydrates
- nucleotides
- lipids and fats

These molecules, however, can have quite a number of molecular variations and combinations, making the world of biological molecules extremely complex. For example, there are over 20 different kinds of biological amino acids, and groups of amino acids often are combined with particular groups of carbohydrates, such as in cell membrane molecules called *glycoproteins* (Figure 1-1).

Each of the first three types of biological molecules can self-combine through covalent bonding to form bioploymers. For example, amino acids combine through peptide bonds to form proteins, carbohydrates combine to form polysaccharides, and nucleotides combine to form nucleic acids.

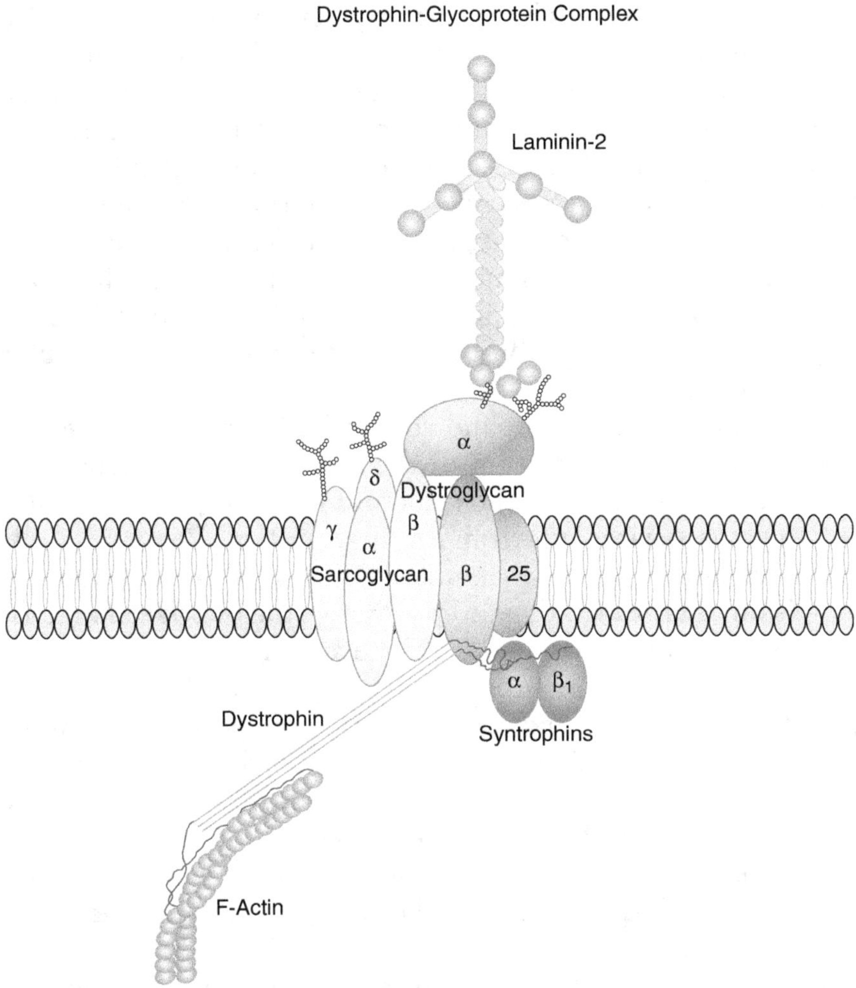

FIGURE 1-1 Dystrophin-glycoprotein complex found in the plasma membrane (sarcolemma) of muscle fibers. The disease Duchenne muscular dystrophy is an X-linked genetic disorder involving the production of the protein dystrophin. Image from Kevin Campbell at the University of Iowa, by permission.

The fourth classification of biological molecules, called *lipids* and *fats,* can undergo self-association, a description of noncovalent interactions, leading to particular "superstructures." For example, cell membranes are composed of associating lipid molecules in a type of structure called a *bilayer*.

Biological molecules have a variety of cellular functions. The biopolymers of amino acids, the peptides and proteins, function as enzymes, cell signal molecules, cellular support structures, and as a cellular energy source. The biopolymers of carbohydrates, the polysaccharides, store energy, function as support structures, and sometimes encode information through particular molecular structure arrangements. Nucleic acids are the primary molecules of information encoding in cells, but they also can act as enzymes and support units. Lipids function as both support structures (membranes) and are an important cellular energy source.

To predict the function and role of biological molecules requires a detailed description of their structure and dynamics. Here we briefly describe the structure of biological molecules and their associated macromolecules, and in Chapters 3, 4, and 5 we address their dynamics. Our focus throughout this book is on "real-time" dynamics—that is, the biological macromolecule dynamics that are actually occurring in the living cell. As we will see, cellular processes represent a well-orchestrated spatial and temporal sequencing of highly specific biomacromolecular interactions in order to carry out their various functions. These interactions are dynamic events that take place over a variety of time scales and length scales. Quantitatively, these interactions often are highly nonlinear, which manifests into a spectrum of behaviors, such as the so-called clock reactions of a cell cycle.

We will also find it useful to adopt a systems perspective of cellular processes. In Chapter 2, we identify five important subsystems of cellular behavior in order to organize the overall functions of the living cell. Within each subsystem, we show how biomolecular interactions dictate the behavior and performance of that particular subsystem. In Chapters 3 and 4, we demonstrate the physical and mathematical approaches necessary to describe the real-time dynamics of these interactions.

In Chapter 5, we demonstrate how the integration of subsystem dynamics allows for a comprehensive analysis of total cellular behavior. Although such a comprehensive analysis is far beyond the reach of current computational simulation capabilities, the necessary approach via the connection of the cellular subsystem dynamics is nonetheless a promising route to the simulation of the living cell.

Chapter 1 provides a minimally necessary description of biological molecular structure for preparation to subsequent discussion and presentation of biomolecular dynamics, which is the central theme of this book. More detailed presentations of biological molecular structure can be found in texts on biochemistry.[1]

Amino Acids, Peptides, and Proteins

The basic molecular formula for amino acids is shown in Figure 1-2. The central carbon atom is called the alpha carbon, and the R group, called the side group, varies depending on the specific type of amino acid. Because of the central carbon tetrahedron, amino acids will exhibit stereo isomerism, resulting in the mirror image forms noted in Figure 1-2. The L-form is by far the most common in nature, but interestingly the D-form does make a rare appearance in some living systems.

Nature has provided us with 20 different forms of amino acids based on the specific side group. The side groups can be charged, polar or nonpolar, leading to a variety of different functions and interactions. The 20 different amino acids are listed in Table 1-1, according to the side group specification.

Amino acids can be covalently linked with one another through so-called peptide bonds, as shown in Figure 1-3, which can result in relatively low-molecular-weight macromolecules with simple linear geometric structures called *peptides* and higher-molecular-weight, more complex geometric structures called *proteins*. In protein synthesis ("translation"), discussed in more detail in Chapter 3, peptide bond formation is catalyzed by the enzyme peptidyl transferase, resulting in rapid production of peptides and proteins by cells.

[1] See, e.g., Pratt and Cornely (2004), and Voet et al. (2006).

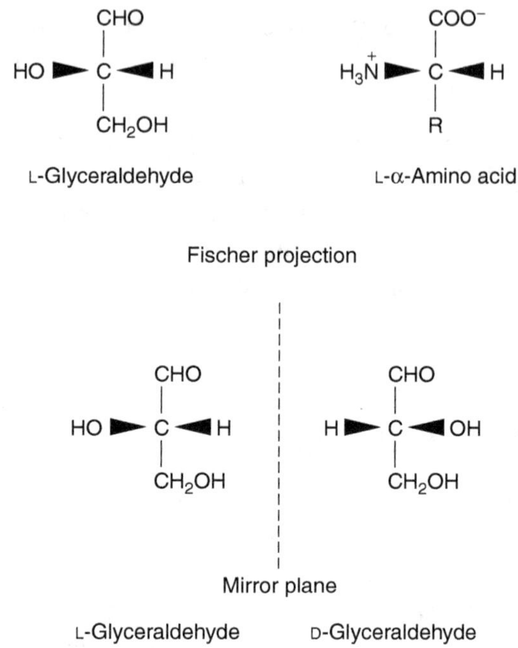

FIGURE 1-2 The basic structure of L-amino acids and comparison with another similar natural molecule, L-glyceraldehyde. The L- and D-forms are mirror images, or stereo isomers, of one another. The D-amino acids rarely are found in nature.

Protein structure is broadly divided into primary, secondary, tertiary, and quaternary classifications. *Primary structure* refers to the particular amino acid sequence in a continuous chain, such as LHLWR representing the chain sequence leucine-histidine-leucine-tryptophan-arginine. *Secondary structure* refers to the most elementary geometric structural patterns that include alpha helixes and beta sheets (see Figure 1-4). *Tertiary protein* structure is the particular geometric arrangement of secondary structures and represents the natural, or folded, state of proteins, as shown in Figure 1-5.

The atomic force interactions of secondary structural units are generally noncovalent with the exception of sulfide bridges that may arise from sulfur containing amino acid side groups. Tertiary structure often is stabilized by protein atom interactions with water. *Quaternary structure* applies only to proteins containing multiple amino acid chains or multiple protein subunits, as exemplified by hemoglobin shown in Figure 1-6. Quaternary

Table 1-1 The 20 naturally occurring amino acids classified according to their particular side chains. The mass is given in daltons. The pK of the side chain is important in establishing its ionization state in solution. For pKs less than the pH (~7) of the solution, the hydronium ions do not remain attached to their associated ionizable group (nonprotonated form). Histidine, with its pK nearly the same as the physiological pH, has an approximately equal molar amount of its protonated and nonprotonated side chains.

Name, Three-Letter Symbol and One-Letter Symbol	Structural Formula	Residue Mass (D)	Average Occurrence in Proteins (%)	pK_1 α-COOH	pK_2 $\alpha\text{-}NH_3$	pK_R Side Chain
Amino acids with nonpolar side chains						
Glycine Gly G	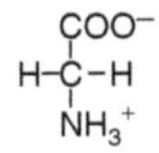	57.0	7.2	2.35	9.78	
Alanine Ala A	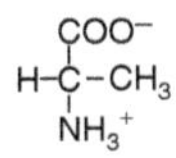	71.1	7.8	2.35	9.87	
Valine Val V	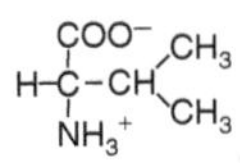	99.1	6.6	2.29	9.74	
Leucine Leu L	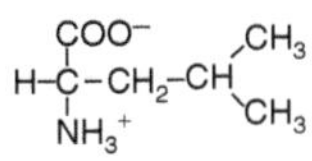	113.2	9.1	2.33	9.74	
Isoleucine Ile I	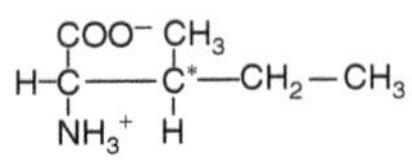	113.2	5.3	2.32	9.76	
Methionine Met M	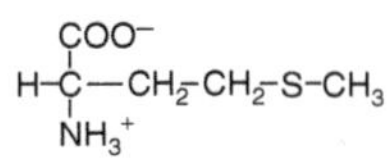	131.2	2.2	2.13	9.28	
Proline Pro P	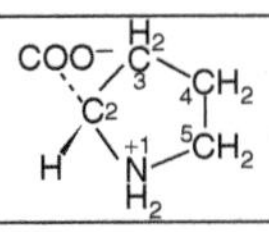	97.1	5.2	1.95	10.64	
Phenylalanine Phe F	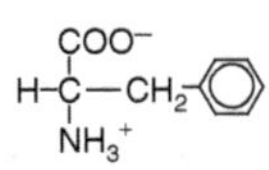	147.2	3.9	2.20	9.31	
Tryptophan Trp W	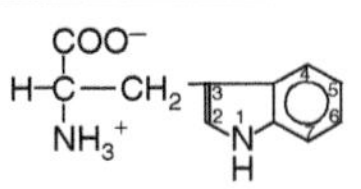	186.2	1.4	2.46	9.41	

*Data from Dawson, R.M.C., Elliott, D.C., Elliott, W.H., and Jones, K.M., *Data for Biochemical Research*, (3d ed.), pp. 1–31, Oxford Science Publications NY (1986).

Table 1-1 (*Continued*)

Name, Three-Letter Symbol and One-Letter Symbol	Structural Formula	Residue Mass (D)	Average Occurrence in Proteins (%)	pK_1 α-COOH	pK_2 α-NH_3	pK_R Side Chain
Amino acids with uncharged polar side chains						
Serine Ser S	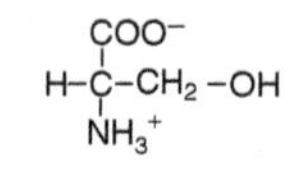	87.1	6.8	2.19	9.21	
Threonine Thr T	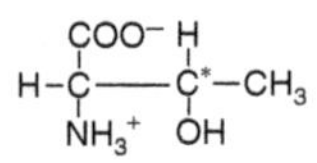	101.1	5.9	2.09	9.10	
Asparagine Asn N	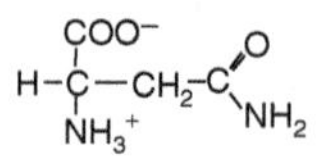	114.1	4.3	2.14	8.72	
Glutamine Gln Q	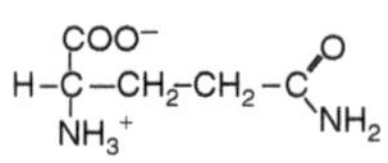	128.1	4.3	2.17	9.13	
Tyrosine Tyr Y	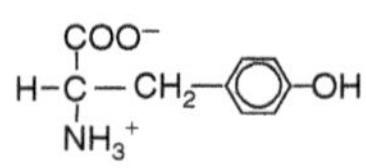	163.2	3.2	2.20	9.21	10.46 (phenol)
Cysteine Cys C	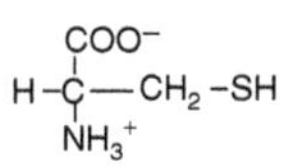	103.1	1.9	1.92	10.70	8.37 (sulfhydryl)
Amino acids with charged polar side chains						
Lysine Lys K	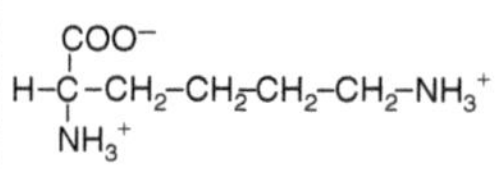	128.2	5.9	2.16	9.06	10.54 (ε-NH_3^+)
Arginine Arg R	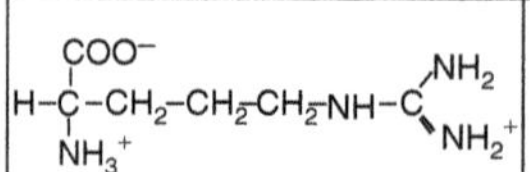	156.2	5.1	1.82	8.99	12.48 (guanidino)
Histidine His H	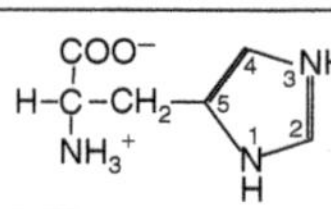	137.1	2.3	1.80	9.33	6.04 (imidazole)
Aspartic acid Asp D	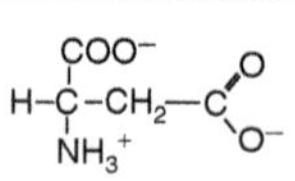	115.1	5.3	1.99	9.90	3.90 (β-COOH)
Glutamic acid Glu E	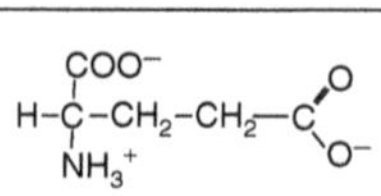	129.1	6.3	2.10	9.47	4.07 (γ-COOH)

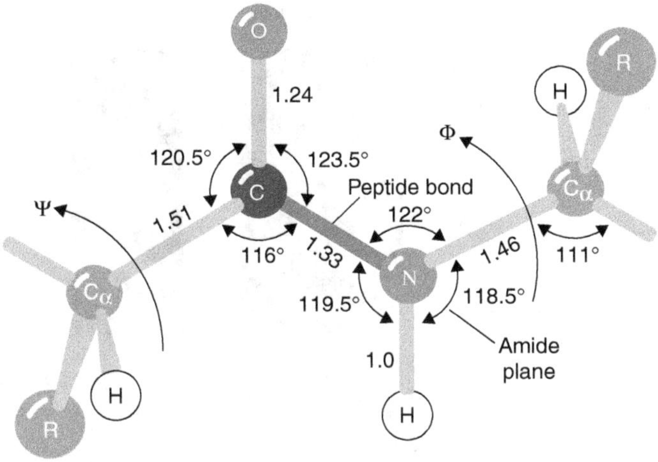

FIGURE 1-3 Illustration of the peptide bond that links amino acids (residues) together. The primary structure of proteins is defined as simply the order of the particular amino acids in a polypeptide chain, beginning at the N-terminal residue. The peptide bond lies in a rigid plane. Also shown are the torsional or dihedral angles around the C_α—N (Φ) and the C_α—C bond (Ψ), which give rise to a multitude of possible conformational states of proteins.

structure helps add geometrical macromolecular symmetry to highly specific protein subunits that generally lack it. The symmetry may help reduce orientational entropic barriers to protein function and still maintain a high degree of specificity via an intact tertiary structure. The natural drawback is the requirement of larger molecules and additional processing steps.

Active proteins are water soluble and typically have charged and polar amino acids on their exterior and more hydrophobic amino acids in their interior. These proteins often are referred to as *globular proteins* (L., *globus*, meaning globe-shaped) because of their compact spherical shape. Proteins can be unfolded or denatured using soaps that disrupt their water soluble structure. Many proteins, such as cell membrane spanning proteins, have both hydrophilic and hydrophobic exterior regions. The hydrophobic portions are found in the membrane region, and the hydrophilic parts are either within the cell (cytosol) or in the exterior fluid (extracellular region).

The wide variety of conformational states of proteins emanates from several fundamental molecular flexibilities

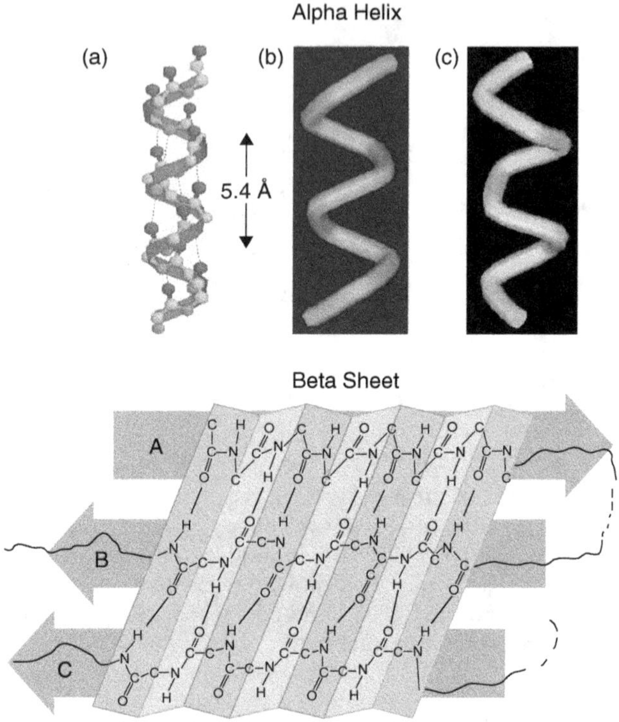

FIGURE 1-4 The two most common secondary structural forms found within polypeptide chains. Alpha helixes can be right-handed or left-handed, and beta sheets can be either parallel or antiparallel.

associated with peptide bonds. As shown in Figure 1-3, peptide bonding between two amino acids results in a rigid plane that is bonded on its vertices by the two associated alphacarbons, the carbonyl oxygen, and the amide hydrogen. The peptide bond (C—N) lies in the center of the rigid plane. Significant molecular rotation, however, around both the $N—C_{\alpha}$ bond and the $C_{\alpha}—C$ bond can take place. The angles associated with these rotations, ϕ and ψ, respectively, are known as either *dihedral* or *torsional angles,* and, as originally shown by Ramachandran, they assume particular, biased values for different secondary structures as shown in Figure 1-7.[2]

[2] See Voet et al. (2006), for more discussion.

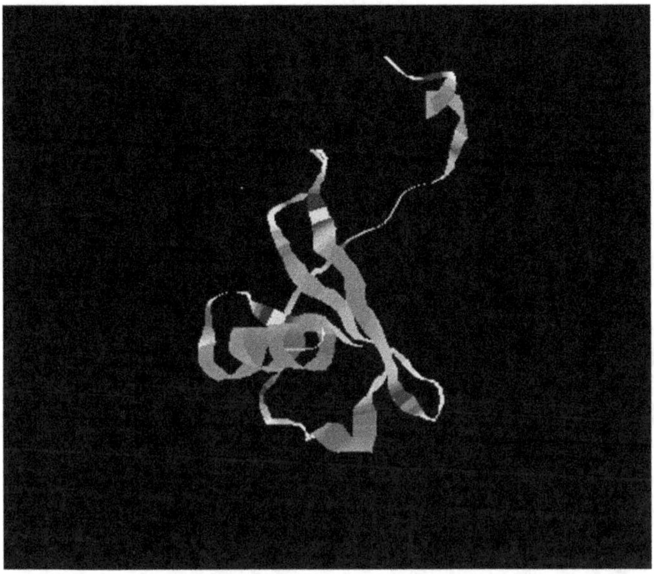

FIGURE 1-5 Illustration of a tertiary, or full three-dimensional, structure of monocyte chemoattractant protein-1 (MCP-1). The polypeptide chain is composed of a number of investing alpha helixes and beta sheets. Illustration created by RASMOL.

The complex structure, multitude of possible conformations, and innate flexibility of proteins enables them to carry out a plethora of cellular activities. Proteins often act in a on/off manner due to the different conformational states any one protein can assume. Proteins, acting as enzymes, contain active sites where cellular biochemical reactions can take place. Protein enzymes can speed up reactions by a millionfold over "bulk" rates.

As an example, chymotrypsin is a protein enzyme secreted by the pancreas to digest proteins from foods into their amino acid constituents. Chymotrypsin, which is always in the "on" state, accomplishes this reduction exercise with a particular active site on its exterior (see Figure 1-8). The active site consists of three amino acids—serine, histidine, and aspartate (catalytic triad)—that break, or cleave, peptide bonds via hydrolysis. The relatively large structure of chymotrypsin, as compared to its small active site, is necessary to produce the exact molecular conformational state of the active site from the folded state of

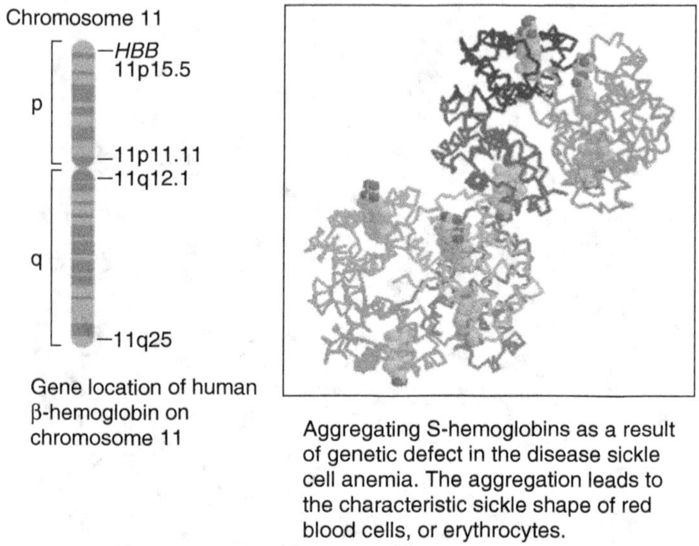

FIGURE 1-6 Example of a quaternary structure in the protein hemoglobin. Here, multiple polypeptide chains are linked together to create a superstructure. Illustrations shown are of aggregating S-hemoglobins from the sickle cell trait. (Protein Data Bank ID: 2HBS; image generated by Protein Explorer.) A summary of the human genome map can be found at the National Library of Medicine site: www.ncbi.nlm.nih.gov.

the protein. The peptide bond region of proteins fit neatly into the catalytic triad on the enzyme surface, weakening the peptide bond's energy and promoting hydrolysis.

Carbohydrates

As shown in Figure 1-9, carbohydrates include the monosaccharides (glucose, fructose, galactose, deoxyribose, and ribose), the disaccharides (sucrose, maltose, and lactose) and the polysaccharides (glycogen, starch, and cellulose).

The monosaccharides are also called the *simple sugars,* and two particular forms—namely, deoxyribose and ribose—are also part of the more complex genetic encoding molecules DNA (deoxyribonucleic acid) and RNA (ribonucleic acid), respectively, as is discussed more fully further in this chapter.

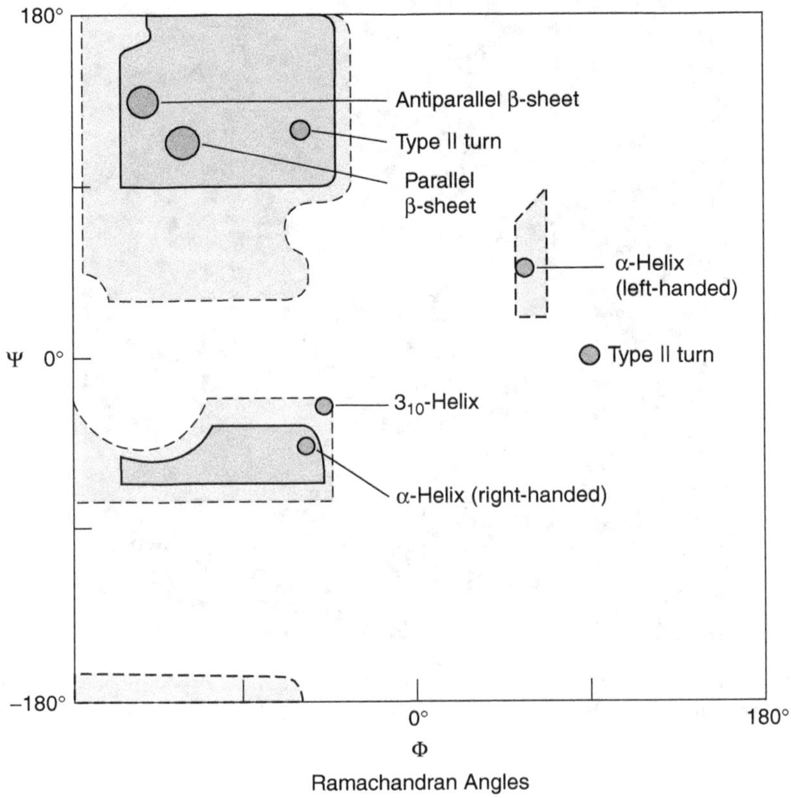

FIGURE 1-7 Ramachandran diagram representing the highest probable occurrences of the peptide backbone angles (Φ, Ψ) shown in Figure 1-3. The "allowed" conformations are the result of steric restrictions that eliminate a majority of all of the *a priori* possible conformational states.

The disaccharides and polysaccharides are formed from dehydration reactions, resulting in ester linkages between the simple sugars (Figure 1-9). For example, *glycogenesis* is the process of producing the polymer glycogen in the liver. The protein enzyme *glycogen synthase* adds glucose molecules together to produce large glycogen molecules for energy storage. Insulin is a protein signal molecule (hormone) secreted by pancreatic β cells that stimulate the process of glycogenesis.

The dehydration reaction to produce glycogen is reversible, and elementary glucose obtained from the polysaccharide glycogen is the important energy molecule of cells. Glycogen is stored

FIGURE 1-8 The protein enzyme chymotrypsin. The highly specific active site is a molecular triad created by the particular folded state of this enzyme. Image generated by RASMOL. (Protein Data Bank ID: Pdb4CHA)

in liver and muscle cells and can be broken down when needed by the process termed *glycogenolysis*. Thus, glycogenolysis, short for glycogen hydrolysis, is the reverse process of glycogenesis (dehydration). The protein enzyme phosphorylase helps breakdown glycogen to glucose. Glucagon (from α cells in the pancreas) and epinephrine (from the nervous system) are cell signal molecules that stimulate the process of glycogenolysis.

Carbohydrate metabolism involves the oxidation of glucose in the mitochondria of cells to produce 36-38 molecules of ATP (adenosine triphosphate), the cells' energy currency, for every single molecule of glucose.

$$Glucose + 6O_2 \rightarrow 6CO_2 + 6H_2O + 36ATP$$

(aerobic oxidation)

The overall process of glucose metabolism is depicted in Figure 1-10. In general, there are three major processes or

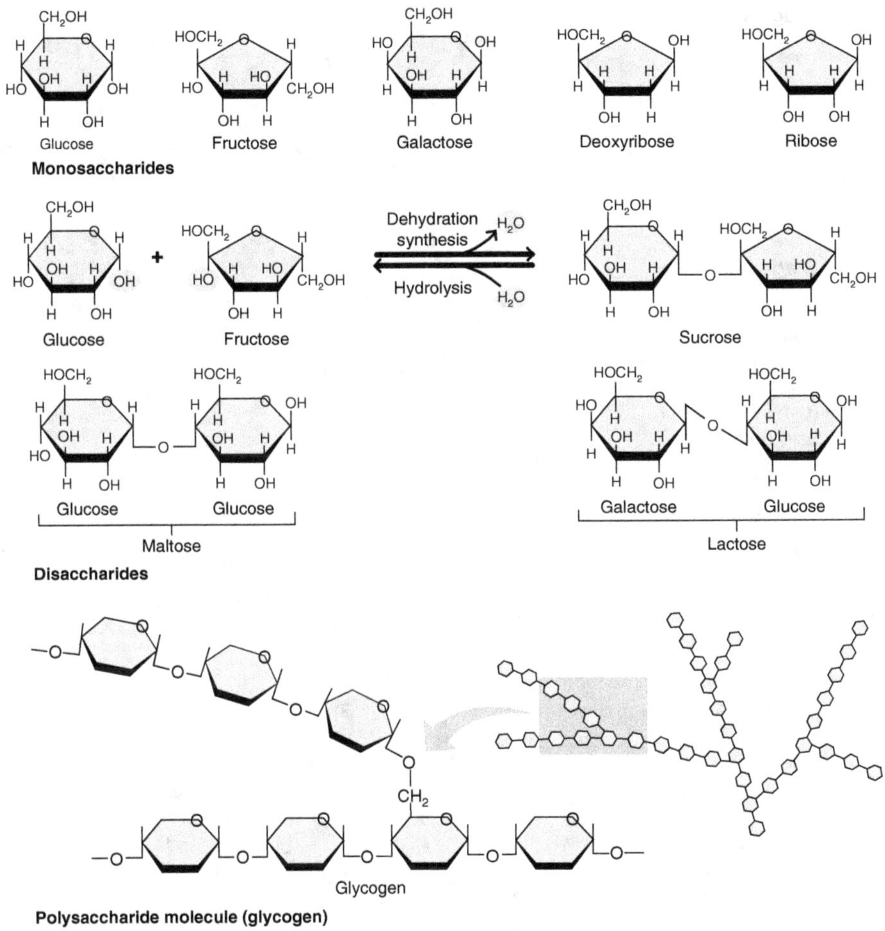

FIGURE 1-9 Carbohydrates and saccharides. Dehydration synthesis reactions allow for the formation of long chains for cellular energy storage. Sequencing of the monosaccharides can be a source of information storage as well.

pathways: (1) glycogenolysis, defined above, (2) glucose oxidation to two and three carbon intermediates within tissue cell cytoplasma, and (3) oxidation of the intermediates in the Citric Acid Cycle, or Krebs cycle, within the mitochondria organelles of tissue cells.

The important carbon intermediate from glucose oxidation is acetyl CoA shown in Figure 1-10. Acetyl CoA enters the Citric Acid Cycle for the final phase of glucose metabolism. However,

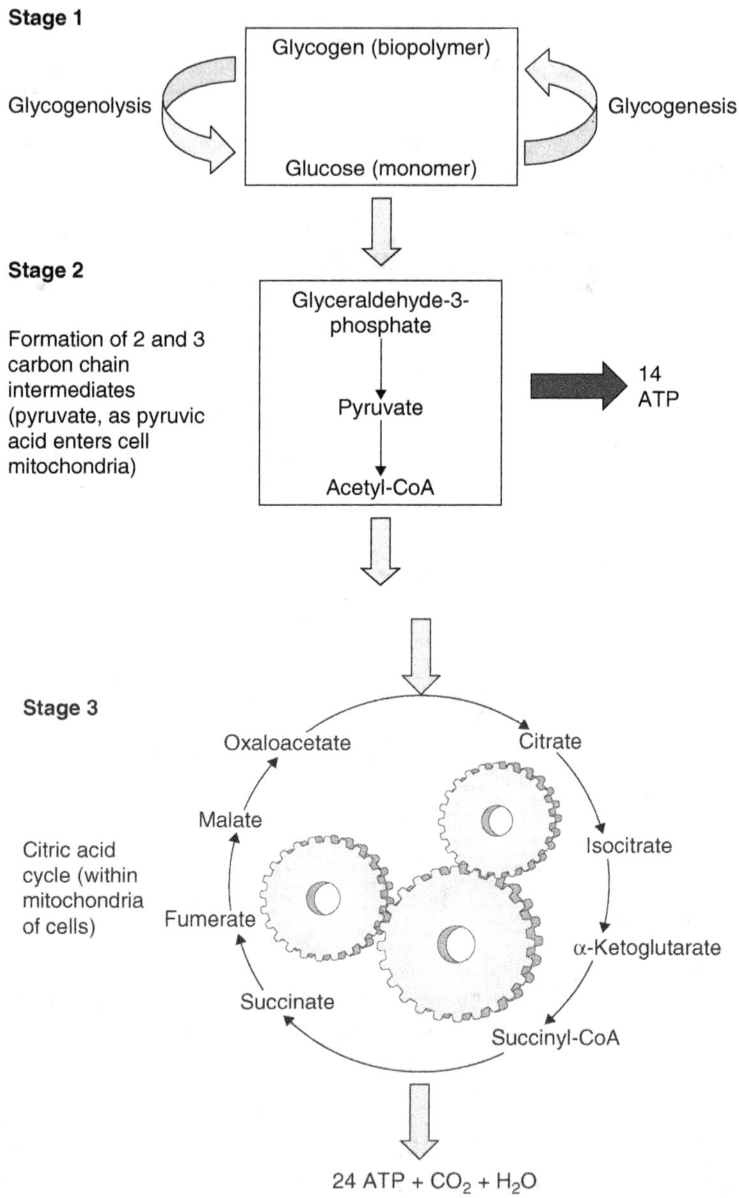

FIGURE 1-10 Overview of glucose metabolism. Other biomolecular metabolic pathways not shown include proteins, lipids and fats, and nucleic acids. Catabolism is the process of breakdown of biological molecules, whereas anabolism is the formation of essential biological molecules.

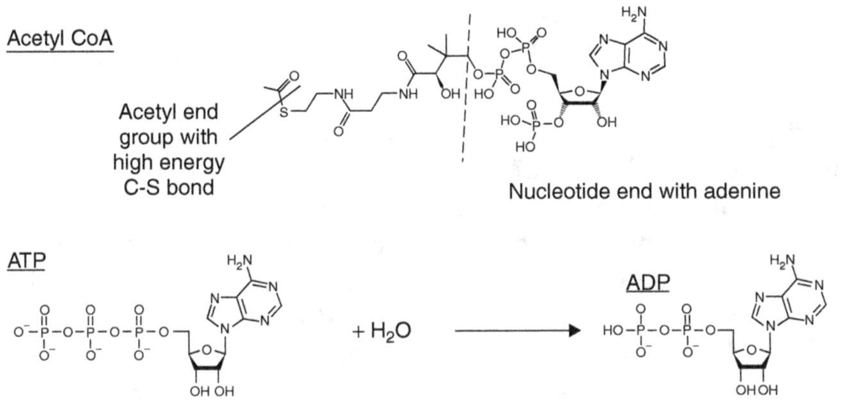

Energetics of the Phosphate Reaction

Reaction	Δ G in kilojoules per mole
$ATP + H_2O \rightarrow ADP + Pi$	–36.8

FIGURE 1-11 Molecular structure of important metabolic molecules acetyl-CoA, ATP, and ADP. The loss of a phosphate group by ATP to from ADP results in significant energy release that is used to "drive" biochemical reactions within cells.

acetyl CoA can also be produced from both amino acid and lipid metabolism, so that proteins and fats can also supply energy via the Citric Acid Cycle when tissue glucose levels are relatively low. Glucose consumption is referred to as catabolism, since the process also works in reverse to generate glucose molecules (anabolic metabolism or anabolism), as well as lipids and fats and non-essential amino acids.

Adenosine triphosphate, or ATP (Figure 1-11), is generated in both the second and third pathway of glucose catabolism. ATP is called the cells' energy currency since it can readily loose a phosphate group to give off needed energy in cellular reaction networks when called upon. Specifically, the reaction and its associated large free energy change are given by

$$ATP + H_2O \rightarrow ADP + P_i, \Delta G^{o'} = -30.5 kJ/mole$$

where ADP is adenosine diphosphate and P_i is a free inorganic phosphate molecule, both resulting from the breakage of the outer phosphodiester bond in ATP. In general, ATP is coupled to biochemical reactions with otherwise positive free energy changes in order to "drive" the reaction or, in other words, to combine into a net negative free energy change for the now coupled reaction system. In this sense, ATP is "money in the bank" for cellular biochemical reactions.

The oxidation reactions of glucose catabolism, like all oxidation and reduction reactions, involve the transfer of electrons (hydrogen atoms) from the oxidized substance to some other molecular specie that is simultaneously reduced ("oxidizing agent"). In glucose catabolism, NAD^+ (nicotinamide adenine dinucleotide) is an oxidizing agent that is reduced, viz.,

$$NAD^+ + H(e^-) \rightarrow NADH$$

Eventually, through numerous oxidation steps, all of the hydrogens are removed from food fuels resulting in CO_2, H_2O, and ATP. NAD^+ must also be regenerated from $NADH$ through a final oxidation step.

The role of simple sugars goes well beyond metabolism in cellular functions. For example, simple sugars are complexed with proteins to form glycoproteins that play important roles in cell structure, immune system molecules, and as cell surface receptor molecules in biochemical signal pathways. As we will see next, sugars are also complexed with the nitrogen base groups, the pyrimidines and purines, in cells to form complex structures called nucleotides.

Nucleotide

Both DNA and RNA are long chains of linear, interconnecting molecules, or residues, called nucleotides (Figure 1-12). Each nucleotide consists of a central sugar group (ribose or deoxyribose), a nitrogen-containing base—adenine (A), guanine (G), cytosine (C), thymine (T), or uracil (U)—and a phosphate group located at the C5′ of the sugar group. The residues are linked by phosphodiester bonds as shown in Figure 1-12 for a segment of

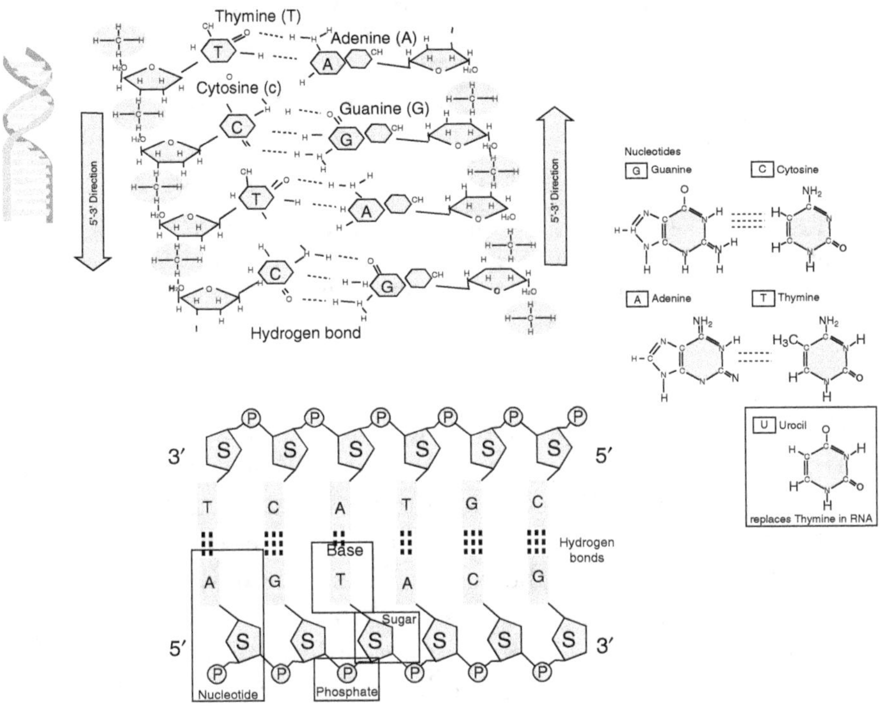

FIGURE 1-12 The basic molecular structure of DNA. Both DNA and RNA are made up of regularly defined geometric complexes of nucleic acids. Nucleic acids are complex molecules formed from three simpler molecular groups—namely, a five-carbon sugar, a phosphate group, and four nucleotides (GCAT in DNA and GCAU in RNA). Unique hydrogen bonding between the nucleotides allows for the storage of information.

DNA. A maximum of four bases exist for either DNA or RNA, because thymine is found in DNA, but not RNA, and uracil is found in RNA, but not DNA.

The order of base groups in a polynucleotide chain is the informational storage system for cellular processes, as detailed in chapters to come. The information or particular ordering is retrievable through specific complementary binding of base pairs via hydrogen bonds. Specifically, the bases of two DNA strands match up according to the Watson-Crick pairing as A-T and G-C, which is illustrated in Figure 1-13. The two strands of complementary DNA form a right-handed, double helix structure

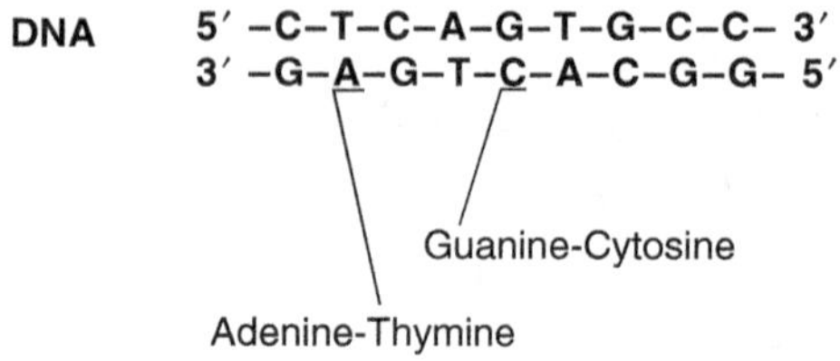

FIGURE 1-13 The unique base pairing in DNA: G-C and A-T. In RNA, uracil replaces thymine. It is the unique base pairing, made possible by hydrogen bonding between the nucleotides, that allows for the storage and transmission of information in the living cell.

in the living cell. The particular geometric features of double-stranded DNA are shown in Figure 1-14. There are 10.5 base pairs per turn (360 degrees), and there is significant rotational flexibility along the helical axis, allowing for the supercoiled configurations found in the cell nucleus. The major and minor grooves allow for sequence recognition by DNA binding proteins, and they help facilitate higher-ordered coiled structures of DNA that allow for compaction of genetic material within cellular compartments.

In transcription, a single strand of RNA known as messenger RNA (mRNA) is produced from a particular open reading frame (ORF) of DNA as depicted in Figure 1-15. Note that a DNA strand and an RNA strand match up by including A(DNA)–U(RNA) as a unique hydrogen bonded structure, since there is no T in RNA. The mRNA sequence is, therefore, unique to the DNA sequence from which it was produced. The 20 different amino acids necessary for particular protein production are made according to the particular mRNA sequence. Every three sequential base pairs defines a corresponding amino acid, as shown in Table 1-2. Because there are four choices for the first base, four choices for the second base, and four choices for the third base, there are 64 (4^3) choices altogether. Thus, there is some redundancy in the base pair ordering for any particular amino acid. Note that two base pairs would lead to only 16 possible combinations and thus be insufficient to code for the necessary 20 different amino acids. Later we describe exactly how proteins are produced from a

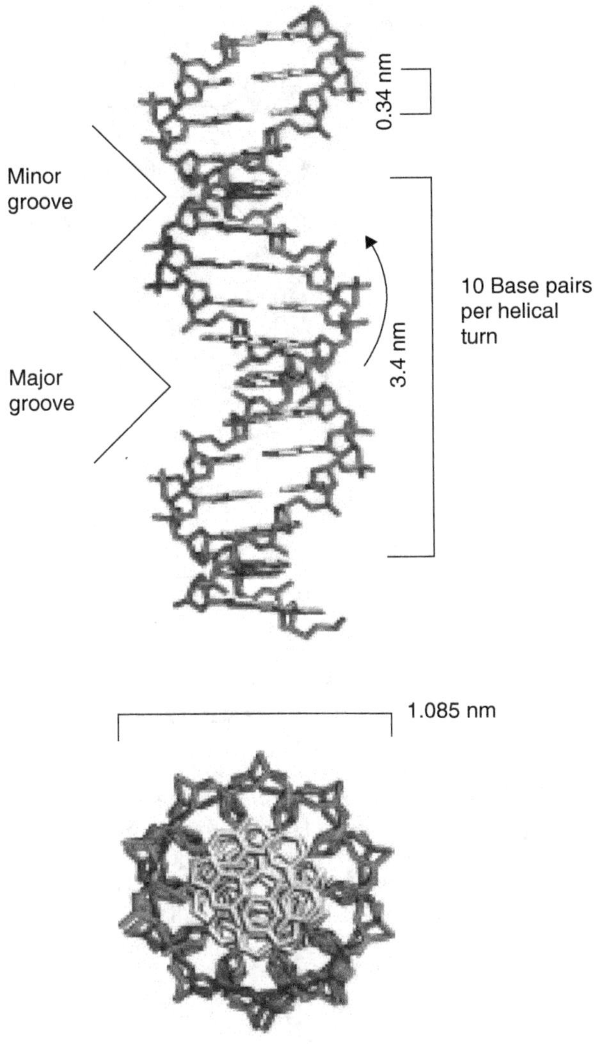

FIGURE 1-14 Structural features of B-DNA, the most common form of DNA. The helical sense is right-handed, and the helical twist per base pair is 31°.

particular mRNA, which is called the *translation process*. Every DNA sequence that codes for a particular protein is known as a *gene*, and in humans there are approximately 35,000 different genes. Posttranslational modifications within the cell then allows for hundreds of thousands of different proteins in the body.

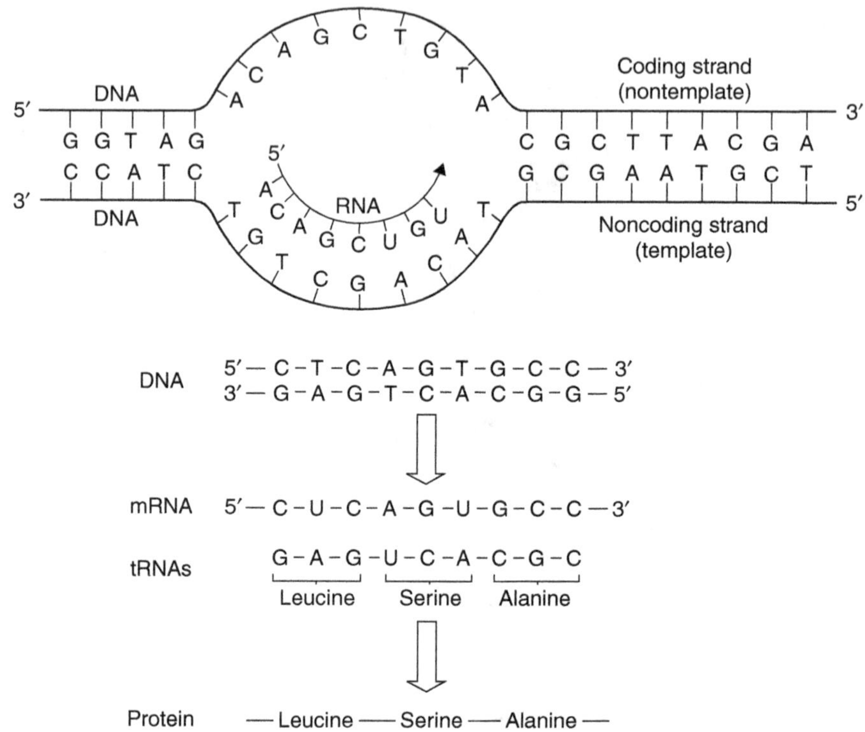

FIGURE 1-15 The overall process of transcription and translation in the living cell. Messenger RNA is produced from the coding segment of DNA via the enzyme RNA polymerase. This first process is called transcription. Proteins are subsequently produced from the mRNA by tRNAs containing specific amino acids for each three-letter nucleoside (the code). This latter process is called translation, discussed in greater detail in Chapter 4. The exact coding of all 20 amino acids is given in Table 1-2.

Lipids and Fats

Fats and oils in biological systems represent various derivatives of triacylglycerol or, simply, triglyceride, as shown in Figure 1-16. Natural fats and oils are hydrophilic in nature and tend to aggregate in aqueous solutions to form fat globules for energy storage and, thus, are an important part of cellular metabolism.

Lipids, on the other hand, are known as fatty acids since they represent long-chain carboxylic acids, as shown in Figure 1-17.

Table 1-2 The standard genetic code. Since there are $4 \cdot 4 \cdot 4 = 64$ choices for the three total positions, there is redundancy in the 20 amino acid assignment code. In addition, UAA, UAG, and UGA all function as stop codons. We will see in Chapter 4 that a particular conjugated molecular form of the AUG sequence functions as a start codon.

The standard genetic code*

First Position (5′ end)	Second Position				Third Position (3′ end)
	U	C	A	G	
U	UUU Phe UUC Phe UUA Leu UUG Leu	UCU Ser UCC Ser UCA Ser UCG Ser	UAU Tyr UAC Tyr UAA Stop UAG Stop	UGU Cys UGC Cys UGA Stop UGG Trp	U C A G
C	CUU Leu CUC Leu CUA Leu CUG Leu	CCU Pro CCC Pro CCA Pro CCG Pro	CAU His CAC His CAA Gln CAG Gln	CGU Arg CGC Arg CGA Arg CGG Arg	U C A G
A	AUU Ile AUC Ile AUA Ile AUG Met	ACU Thr ACC Thr ACA Thr ACG Thr	AAU Asn AAC Asn AAA Lys AAG Lys	AGU Ser AGC Ser AGA Arg AGG Arg	U C A G
G	GUU Val GUC Val GUA Val GUG Val	GCU Ala GCC Ala GCA Ala GCG Ala	GAU Asp GAC Asp GAA Glu GAG Glu	GGU Gly GGC Gly GGA Gly GGG Gly	U C A G

*The 20 amino acids are abbreviated: Ala, alanine; Arg, arginine; Asn, asparagine; Asp, aspartate; Cys, cysteine; Gly, glycine; Gln, glutamine; Glu, glutamate; His, histidine; Ile, isoleucine; Leu, leucine; Lys, lysine; Met, methionine; Phe, phenylalanine; Pro, proline; Ser, serine; Thr, threonine; Trp, tryptophan; Tyr, tyrosine; and Val, valine.

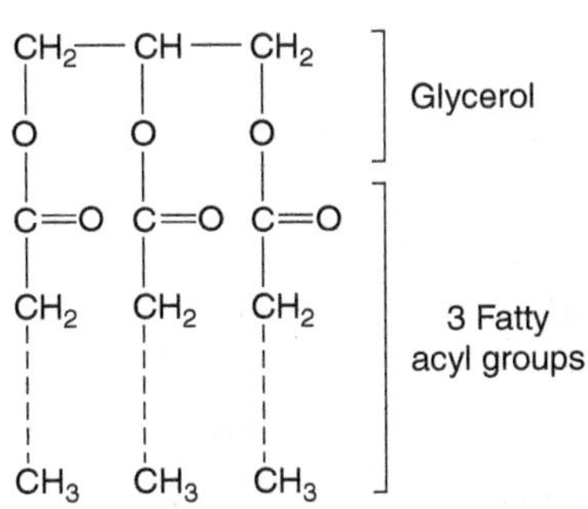

FIGURE 1-16 Basic molecular structure of triglyceride or triacylglycerol. These molecules represent a critical energy source in animals, and they are synthesized and stored in adipocytes (fat cells).

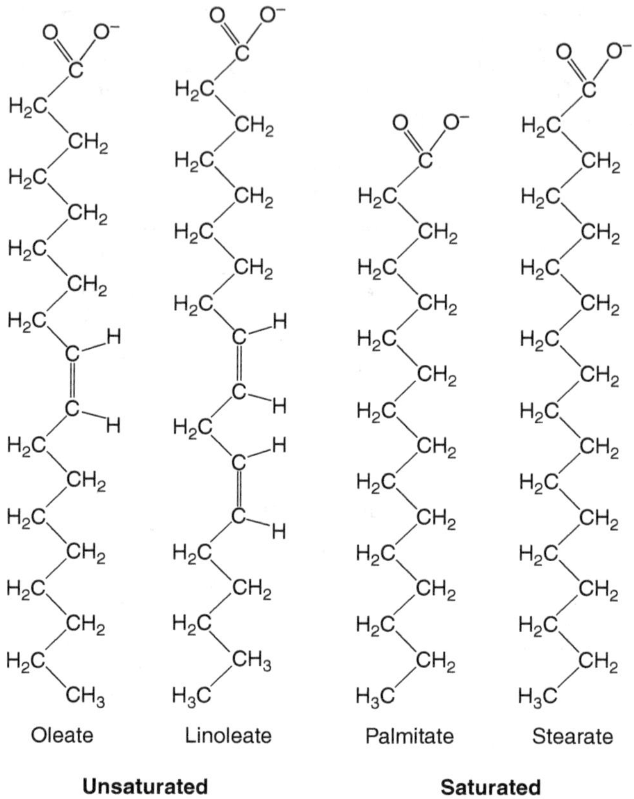

FIGURE 1-17 Basic molecular structure of lipids. These molecules are long-chain fatty acids. The carbon tails can contain all saturated carbon atoms (saturated fatty acids) or some unsaturated carbon atoms (unsaturated fatty acids).

Unsaturated fatty acids contain at least one carbon double bound in the chain, as illustrated in linoleate in Figure 1-17, whereas saturated fatty acids contain no double bonds, such as palmitate shown in the same figure.

An important class of lipids are the glycerophospholipids that make up cell membranes (Figure 1-18). The phospho-head group is highly charged and hydrophilic, whereas the long chain carbon-tail group is electrically neutral and hydrophobic. As a consequence of their amphiphilic nature, these molecules can structurally form bilayers in aqueous solutions (Figure 1-19). As cellular membranes, these bilayers physically encapsulate

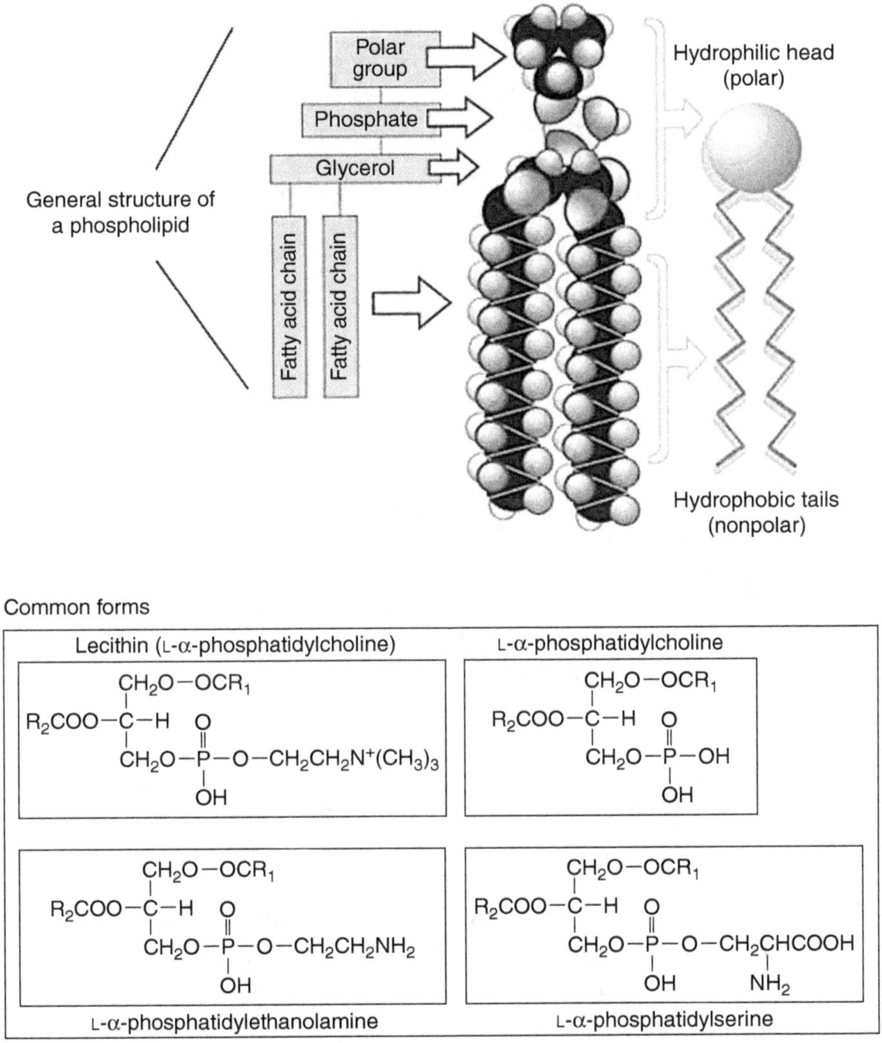

FIGURE 1-18 The basic molecular structure of phospholipids. These molecules are called amphiphilic because they contain a hydrophilic head group (phosphate group) and a hydrophobic carbon-tail group. This double role results in a variety of possible macromolecular structures, depending on the surrounding fluid and specific intermolecular interactions. One important example is the bilayer shown in Figure 1-16.

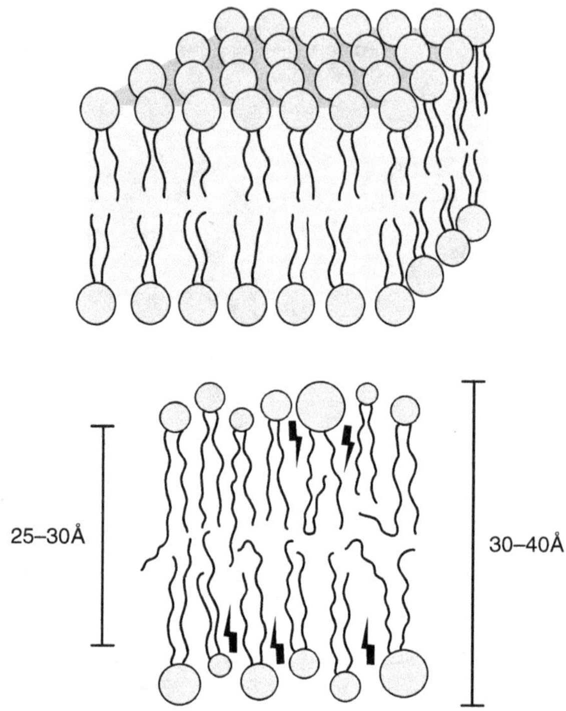

FIGURE 1-19 Illustration of the basic structure of a lipid bilayer making up the cell membrane. The small dark-shaded molecules in the lower picture represent cholesterol, which helps stabilize the bilayer and provide more rigidity to the membrane.

cellular contents. Cellular membranes are sometimes referred to as being fluid in that the lipids can undergo a planar-type motion within the bilayer structure. The cellular bilayer contains cholesterol (a single-chain fatty acid) for improved structural rigidity and a variety of membrane molecules for communicating with the extracellular matrix. These latter molecules include integral membrane proteins for cell signaling, glycoproteins for self-identification, and a variety of channel proteins to allow for the transport of molecules into and out of the cell proper.

Finally, we also include within the lipids and fats category compounds called steroids and *hormones* which are important cell signal molecules mainly derived from cholesterol (a fatty acid). These compounds can either up-regulate or down-regulate cellular functions, including metabolism and reproduction.

Solvent Composition and Molecular Structure Information

A critically important aspect of cellular behavior is the solvent composition itself. The cellular solvent, excluding macromolecules, contains a number of charged atoms and molecules, including hydronium ions H^+, hydroxy ions OH^-, salts, such as K^+, Na^+, Cl^-, and buffering components (weak acids and their conjugate bases). These charged atomic groups can greatly affect the interactions and behavior of biological macromolecules, as discussed in more detail in subsequent chapters. A typical solvent composition of a eukaryotic cell cytoplasm is shown in Table 1-3.[3]

In addition to the quantitative characterization of the solvent, the specific molecular structure of cellular molecules is required to understand cellular behavior. The molecular structural features of proteins, mRNA, DNA, and so on dictate each molecule's interaction and specificity within the cellular world. Molecular structural information can be obtained by a variety of methods, such as nuclear magnetic resonance (NMR) imaging, x-ray crystallography, atomic force microscopy, and so on.[4] No attempt is made here to review all of the particular experimental methods, and we simply refer the reader to the references at the end of this chapter. There continues to be improvement in our computational ability to predict molecular structure by *ab initio* methods in the field of quantum mechanics.[5] Molecular structure information, from either experimental or computational methods, is a required starting element of everything shown here in the analysis of the real-time behavior of biological macromolecules and their interactions.

---

[3] For a complete description of cellular systems see Alberts et al. (2002, 2004).

[4] See, e.g., Hammes (2005).

[5] See, e.g., Leach (2001).

Table 1-3 The overall composition by volume of the living cell. The cellular contents outside of water, the macromolecules, phospholipids, ions, and small molecules carry significant electrical charge that helps binds water inside the cell. Hydronium ion concentrations inside the cell are slightly higher than the extracellular fluid, giving the cell a relatively lower pH. In addition, there exists a slightly more net negative charge inside the cell as compared to the extracellular fluid. This gives a net negative electrical potential difference across the plasma membrane (resting potential), and its value varies from about −60 mV to −90 mV, depending on the specific cell type.

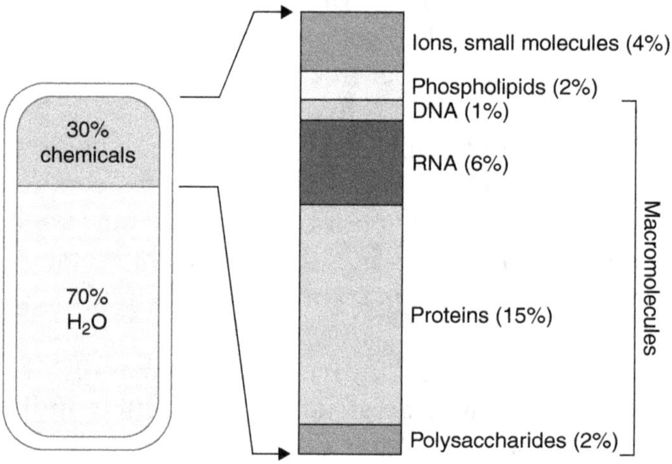

Salt concentrations

Intracellular	Extracellular
Na^+ 27 mM	Na^+ 145 mM
K^+ 123 mM	K^+ 5 mM
Cl^- 6 mM	Cl^- 125 mM
(pH ≅ 6.8)	(pH ≅ 7.4)

References

Alberts, B., Bray, D., Hopkin, K., Johnson, A., Lewis, J., Raff, M., Roberts, K., and Walter, P., *Molecular Biology of the Cell,* 4th ed., Garland, New York (2002).

Alberts, B., Bray, D., Hopkin, K., Johnson, A., Lewis, J., Raff, M., Roberts, K., and Walter, P., *Essential Cell Biology,* 2d ed., Garland, New York (2004).

Hammes, G.G., *Spectroscopy for the Biological Sciences,* Wiley, New York 2005.

Leach, A., *Molecular Modeling. Principles and Applications,* Prentice Hall, New Jersey (2001).

Pratt, C.W., and Cornely, K., *Essential Biochemistry,* Wiley, New York (2004).

Voet, D., Voet, J.G., and Pratt, C.W., *Fundamentals of Biochemistry,* 2d ed., Wiley, New York (2006).

Chapter Two

Cellular Systems Biology

Functionally, we can divide cellular processes into five different subsystems, as depicted in Figure 2-1. These systems are (1) external/internal communication and transport pathways (including membrane transport); (2) enzymatic and biochemical production; (3) cellular reproduction or division; (4) mechanical work, motility (taxis), and structural support; and (5) information systems, checkpointing/safety, and reliability (genotyping and phenotyping).

These systems interact with one another and do not function independently, and their interactions often are precisely coordinated both in time and space. It is necessary to describe each subsystem both qualitatively and quantitatively in order to begin to integrate subsystem behavior into an overall prediction of cellular processes. So, we begin by briefly discussing each

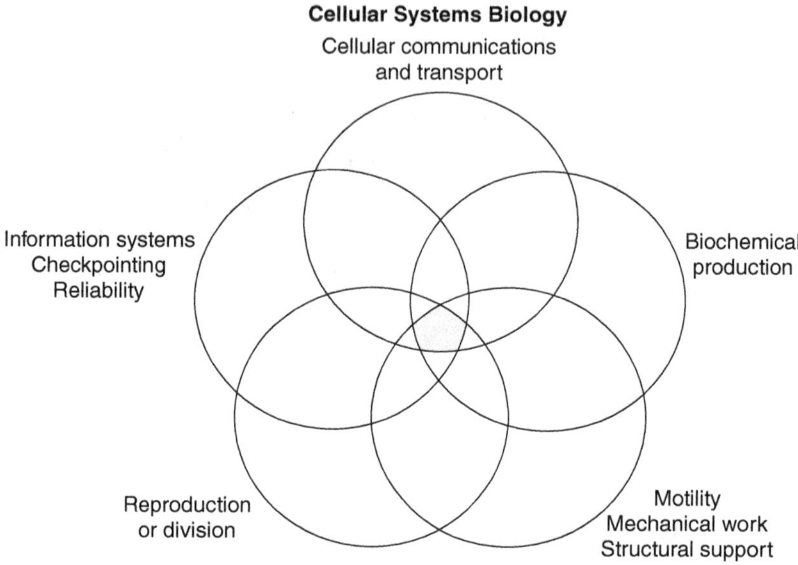

FIGURE 2-1 Venn diagram representation of the five cellular subsystems. These subsystems interact in complex ways to accomplish overall cellular processes.

subsystem qualitatively and then conclude this chapter with some elementary quantitative aspects of length and time scales. Chapters 3, 4, and 5 examine the detailed quantitative, dynamic aspects of biomacromolecular behavior that provide a basic starting point for rigorously quantifying cellular subsystem behavior.

External/Internal Communication and Transport Pathways

In general, cells communicate through the use of chemical messenger or signal molecules. Cytokines, for example, are the messenger molecules of the immune system. Hormones are messenger molecules that up-regulate or down-regulate metabolic and growth processes. We can functionally classify cell signaling and communication pathways as either extracellular or intracellular. Signal molecules may also initiate the opening of channels in cell membranes, allowing for the transport of materials into and out of the cell. Biochemical transport pathways, including extracellular, intracellular, and membrane, often limit the overall rates of cellular processes.

Extracellular chemical signaling may be further divided into (1) local, or proximal, signaling and (2) distal signaling.[1] Local signaling includes paracrine signaling, neuronal signaling, and contact signaling. In local signaling, molecules are released by nearby cells traveling through the extracellular matrix to find their target cell receptors. For example, epidermal growth factor (EGF) and platelet derived growth factor (PDGF) are cellular signals released in response to injury, or insult, to local tissue. These paracrine signal molecules find their local target receptor molecules (EGFR, PDGFR, etc.), such as displayed on the surface of mast cells and epithelial cells, that help launch a coordinated set of countermeasures against the insult. Neuronal signaling is another example of local signaling whereby neurotransmitter molecules, such as acetylcholine, are released by neurons at synapses or at neuromuscular junctions. Contact signaling between cells helps maintain a proper differentiated state for the overall integrity of the associated tissue.

Endocrine or distal signaling is a systemic pathway chiefly associated with hormones, such as insulin, glucagon, thyroid hormone, estradiol, testosterone, and so on. Endocrine signal molecules move through systemic pathways, such as vascular and lymphatic pathways, to reach their target receptors. Norepinephrine, for example, is released by the adrenal glands to the blood and helps provide a sympathetic autonomic response at various target sites throughout the body, such as cardiomyocytes (increase heart rate), epithelial cells of the digestive tract (inhibit digestive processes), and so on.

Once signaling has been initiated through the extracellular pathway, further signal transmission or cascades usually are required to initiate a particular cellular response, such as through so-called second messenger systems (G-protein linked receptors). In addition, internal cell communication molecules are required to help maintain or modulate basic cellular activity

[1] See, e.g., Alberts et al. (2003).

(homeostasis). Many internal signaling processes represent amplifications of external, initiating signals. Amplifications are necessary to illicit specific cellular processes because such processes often are limited by intracellular species transport, diffusion, and entropic barriers that may be overcome at higher species concentrations. Modulation often is necessary to avoid an overresponse, and the entire process can be quantified through control loop structure and analysis, as discussed further in Chapter 5.[2]

Membrane transport also is a communication link between cells and the extracellular matrix, although it often is considered as a separate cellular subsystem. Intracellular membrane transport also occurs in eukaryotic cells across nuclear membranes and, generally, across membrane-containing cell organelles. Membrane transport can take place via passive (Fickian) diffusion of very small hydrophobic molecules through the lipid bilayer or both passively and actively through membrane protein channels or pores. Many of these protein channels can assume open or closed positions to regulate the flow of materials into and out of cells.[3] Channels open or close in response to signal molecules or to extracellular or intracellular chemical concentration gradients, such as ionic species gradients (Figure 2-2). Sodium and potassium channels in neurons, for example, propagate action potentials created by some initial ionic or chemical species concentration gradient.

In general, there are numerous types of intracellular and extracellular transport pathways. Intracellular highways for rapid biomolecular transport are the result of intracellular spatial construction and order, as is further discussed in Chapter 5. Extracellular transport pathways are created by the particular cell tissue matrix, including gap junctions, void fractions, membrane potentials, and so on.

In general, the overall rates of cellular processes may be limited by biomolecular transport rates, including external and internal communication pathways. Chapters 3 and 4 discuss

---

[2] For more details, see Gomperts (2003), and Bradshaw and Dennis (2003).

[3] See, e.g., Hille (2001).

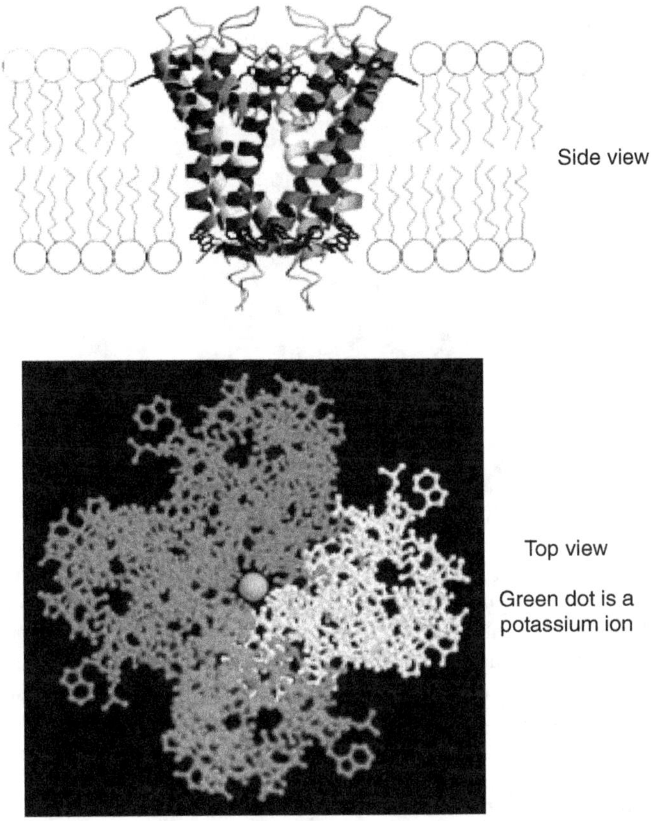

FIGURE 2-2 Molecular structure of a voltage-gated potassium membrane channel. The voltage-gated potassium channel shown here is one of the few channels where a complete three-dimensional molecular structure has been obtained. (Ref.: R. MacKinnon et al. *Science,* 309, 897, 2005.)

how the field of scaling and dimensional analysis can provide much insight into these limitations.

Enzymatic and Biochemical Production Systems

The living cell must constantly produce new molecules for all of its various functions. Some of these molecules are produced through the catabolic and anabolic pathways of metabolism. Proteins are produced through transcription and translation processes, described in detail later, and a variety of biochemical

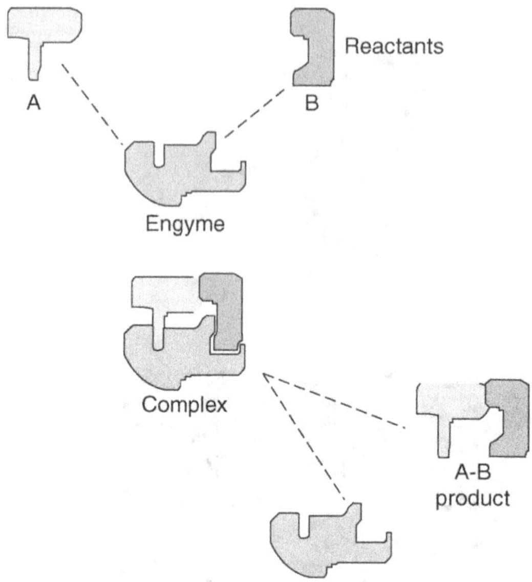

FIGURE 2-3 Simple enzymatic reaction scheme. The enzyme's role is to bring reactants A and B together in the right orientation and energy for the reaction to proceed. Enzymes, however, can function in more complex ways. For example, enzymes can assume multiple conformations, such as active and nonactive states, that can be regulated by other cell molecules and downstream reaction products.

compounds are produced from precursor molecules via enzymes. Enzymes play a key role in speeding up otherwise immeasurably slow chemical reactions. The rates of enzyme production and removal are, therefore, an excellent regulator of overall biochemical production processes.

In general, enzymes are catalysts that greatly speed up chemical reactions, but are not themselves products of the reaction.[4] Enzymes work as catalysts by, for example, bringing two molecules (A + B) together in local binding site regions on their surface or within cleft regions (Figure 2-3). As a result of their close proximity and proper relative orientation, the two molecules subsequently undergo covalent bonding to produce a product molecule. Without the enzyme surface, the odds of a random encounter of A and B in just the correct six-dimensional orientation (three orientation degrees of freedom

[4] For a more complete discussion, see Voet et al. (2005).

for each macromolecule) are highly unlikely, and biochemical reaction rates in the absence of enzymes are usually so small as to be immeasurable.

Enzymes are biomacromolecules, typically proteins, that work in many interesting ways—for example, so-called allosteric enzymes can adopt two different conformational states due to the enzymes' internal flexibility. One confirmation may represent an active state, whereby the enzyme promotes the reaction, and a second enzyme conformation may represent the inactive, or noncatalytic, state. Regulatory molecules allow for the switching of conformational states and thereby act as a chemical control mechanism for biochemical production. Regulatory molecules may derive from external and internal communication pathways. In a series of numerous enzymatic reactions, the downstream products themselves often act as the regulatory molecules—in this case, inhibiting further chemical production at the front end; this is called *feedback inhibition,* or *negative regulation.* Positive regulation, on the other hand, can speed up front end production by activating more enzymes.

Clearly, to understand the complexities of biochemical production processes requires a sophisticated quantitative analysis of the reaction pathways. Feedback loops introduce complex nonlinear behavior that can have important consequences to cellular function. Important cellular patterns, such as oscillatory biochemical production behavior and the so-called clock reactions that set cell cycles, are a consequence of the high degree of nonlinearity of underlying reaction mechanisms (Figure 2-4).[5]

Reproduction or Division Systems

Most cells of the over 200 cell types in our body exist in a final differentiated state and do not undergo cell division or mitosis. Those cells, however, typically wear out and must be continuously replaced, as with the epidermal cells of our skin. In humans, the entire outer surface, or epidermal layer, of our

[5] Heinrich and Schuster (1996).

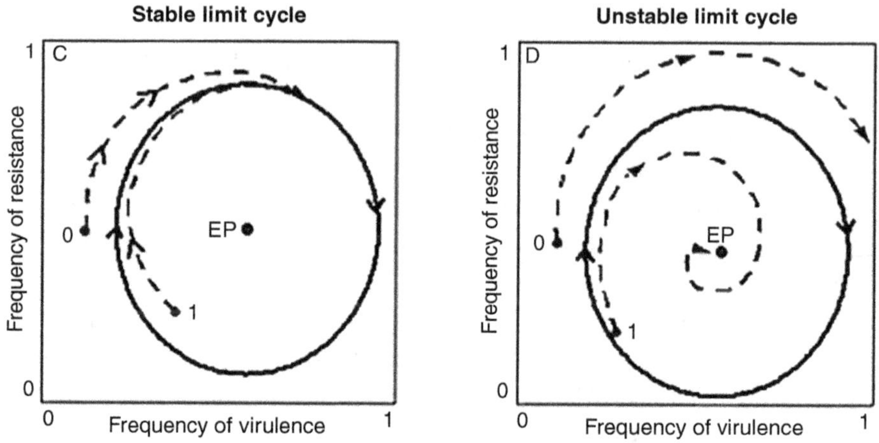

FIGURE 2-4 Illustration of limit cycles in plant pathogen systems. Here the particular frequencies of resistance and virulence genes lead to different limit cycle behaviors. Reproduced by permission from the American Phytopathological Society.

skin (or integumentary system) is replenished every few months, so the face of the person that you see now will be completely replaced in a few months' time! Stem cells are the precursor cells that undergo mitosis or cell division to produce new, differentiated cells. For example, bone marrow stem cells produce new blood cells (erythrocytes and leukocytes) continually as the differentiated blood cells wear out. It is interesting to note that oftentimes the components of worn-out cells are recycled and used in some fashion by the body. For example, the worn-out erythrocytes just mentioned are transformed to bilirubin in the spleen and transported to the liver to be used as bile in digestion. Proteins from worn-out body cells, such as GI tract epithelial cells, are broken down to their elementary amino acids to be used in making new proteins.[6]

Stem cells have special machinery for reproduction or cell division in terms of various proteins and transcription factors needed throughout the complicated process of reproduction. These factors are the result of particular gene expressions that

[6] We could call this a recycle system that would be the eco-friendly division of the biochemical production system just discussed.

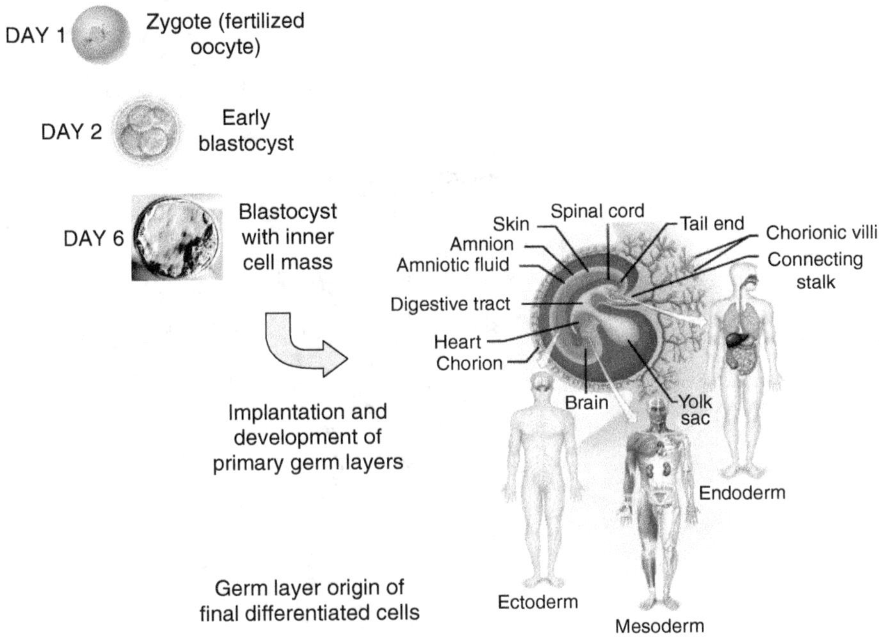

FIGURE 2-5 Embryonic cellular origin of human tissue. The primary germ layers—ectoderm, mesoderm, and endoderm—give rise to the different human tissue types. Part of this figure was reproduced by permission from the McGraw-Hill Companies.

are silenced in differentiated cells. Gene expression and the presence of various transcription proteins are, therefore, important factors in the classification of stem cells. Embryonic stem cells, derived from blastocysts, are so-called *totipotent* in that they are capable of differentiating into every type of body cell. Other stem cells are called *pluripotent*—that is, capable of producing several kinds of differentiated cells. The different types of body cells and their embryonic origins are illustrated in Figure 2-5.

The general cycle of cell division is shown in Figure 2-6.[7] The first major phase of division is known as the *interphase* and consists of three subphases—that is, a G_1 (initial growth) phase, an S (DNA synthesis) phase, and a G_2 (final growth) phase. In brief, the interphase does all of the necessary biochemical production

[7] For a more complete presentation of developmental biology, see Gilbert (2003).

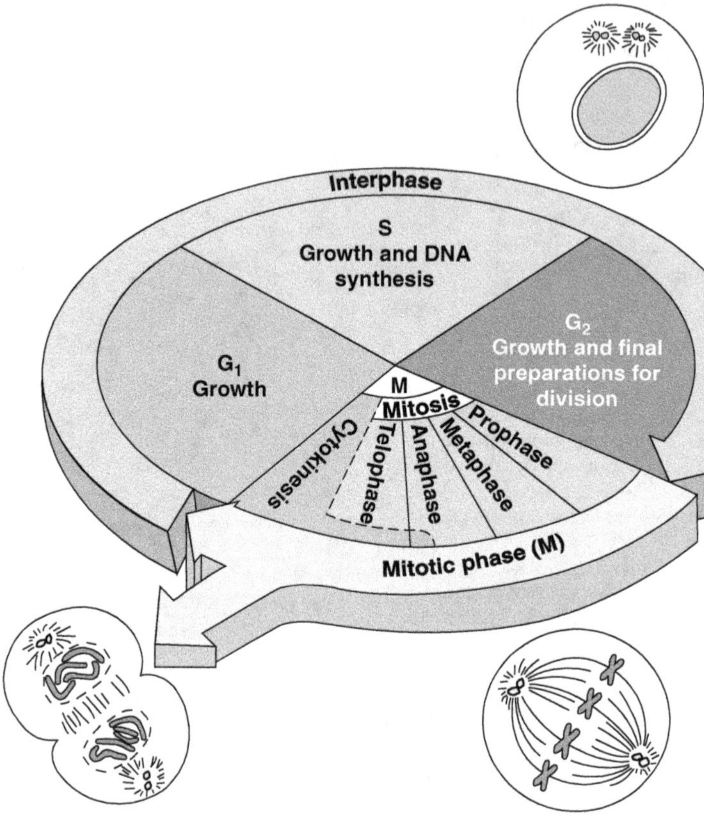

FIGURE 2-6 General cycle of cell division showing the major phases, interphase and mitotic, and their associated subphases. Reproduction from Elaine M. Marieb, *Human Anatomy and Physiology,* Pearson Publishing, by permission.

needed for the final, more mechanical phase called the *mitotic phase*. In the mitotic phase, a number of different micro-visually identifiable stages occur, as shown in Figure 2-6. Note that differentiated cells are arrested outside of the G_1 phase in what is called the G_0 phase. Cancer cells are differentiated cells that have entered the cell division cycle through the occurrence of aberrant genes.

In general, cell division is classified as either mitosis or meiosis. In *mitosis,* a diploid cell (a cell with maternal and paternal chromosome pairs) divides to produce two identical diploid cells. In *meiosis,* a gamete diploid cell divides twice to produce four haploid cells (cells with unpaired chromosomes). In humans, meiosis is the process of producing sperm in males and oocytes in females. A comparison of mitosis to meiosis is shown in Figure 2-7.

It is interesting to note that mature oocytes contain all of the biochemical constituents, except the complete set of genetic material, necessary for cell division. This includes storages of transfer RNA (tRNA), amino acids, carbohydrates, lipids, and so on. Modern nuclear reprogramming and cloning methods involve the simultaneous removal of the oocyte haploid DNA and insertion of somatic cell diploid DNA from a donor. Through various external stimulation methods, the reprogrammed oocyte, or what is now a sort of prezygote, is coaxed to divide and become a source of totipotent stem cells. Such approaches, although still in their scientific and technical infancy of development,

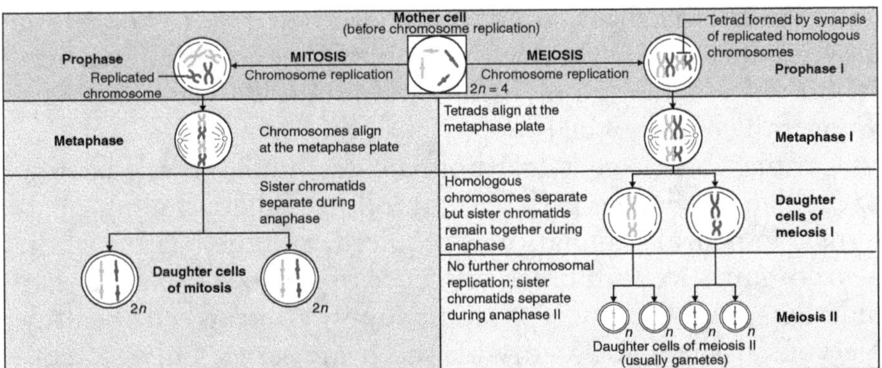

Event	Mitosis	Meiosis
Number of divisions	One; consisting of prophase, metaphase, anaphase, and telophase	Two; each consisting of prophase, metaphase, anaphase, and telophase; DNA replication does not occur between the two nuclear divisions; an event unique to meiosis is that during meiosis I, the homologous chromosomes synapse (join along their length), forming tetrads (groups of four chromatids)
Number of daughter cells and genetic composition	Two; each diploid ($2n$) and identical to the mother cell	Four; each containing half as many chromosomes as the mother cell (hence haploid or n); nonidentical to mother cell
Importance in the body	Development of multicellular adult from Zygote; to produce cells for growth and tissue repair; ensures constancy of genetic makeup of all body cells	To produce cells for reproduction (gametes); introduces genetic variability in the gametes and reduces chromosomal number by half so that when fertilization occurs, the normal diploid chromosomal number is restored (in humans, $2n = 46$)

FIGURE 2-7 Comparison of mitosis and meiosis. Mitosis results in the formation of two identical, or cloned, diploid cells from a single diploid cell, whereas meiosis results in the formation of four haploid cells beginning with a single diploid cell. Chromosomal cross-over provides genetic variation in the process of meiosis. Reproduction from Elaine N. Marieb, *Human anatomy and physiology*, Pearson Publishing, by permission.

allow for the production of any type of autologous body cell for therapeutic applications.[8]

Mechanical Work, Motility Systems, and Structural Support Systems

Motility is the bulk motion of a cell as a whole, and, therefore, a different type of subsystem from those associated with the "static" cell just discussed. Not all cells have the property of motility, and motility may be suddenly expressed by a static cell as in cancer metastasis described here. Within this functional subsystem, we also include muscle cells that change their shape to do mechanical work.

Perhaps the most studied and well-known example of cell movements is in the movement of skeletal muscle fibers that actually are muscle cells. These cells are stimulated via motor nerves to exhibit short, rapid, and coordinated movements. The movements actually are due to complex interactions of particular proteins (actin and myosin) within muscle fibers that cause an overall change of cell shape.

Cardiac muscle cells (cardiomyocytes) undergo self-beating or oscillatory changes in shape that follow regular, rhythmic patterns. No neural stimulation is required, and the beating of the heart organ is a coordinated sequence of events, via cellular gap junctions, requiring only a blood supply to keep cells healthy. Nervous tissue associated with the heart serves only to regulate heartbeat—specifically, parasympathetic vagal nerves slow down the natural rate, and sympathetic spinal nerves speed the heart rate up (tachycardia).

As opposed to shape changes that do mechanical work, bulk cell motion involves a migration of cells in response to an external stimulus. For example, cells of the immune system, such as macrophages, are drawn to an "injured" area in response to chemical signals. *Chemotaxis* is a term used to describe cell

[8] Kiessling and Anderson (2003).

movement along extracellular chemical concentration gradients. Other cells, such as fibroblasts, can undergo a sort of walking motion through the extracellular matrix. They extend a group of adhesion fibers (principally actin) via a polymerization reaction in the forward direction of cell movement. Nearly simultaneously, adhesion fibers are retracted in reverse direction of motion. Judiciously placed adhesion receptor molecules in the extracellular matrix lead to a guided cell movement.[9]

Static cells remain attached to one another, or to the extracellular matrix, via a complex set of membrane-spanning molecules called *integrins*. The expression of integrins by cells is, therefore, an important identifier of the state of cell motility.

In metastasis, cancer cells can spread by actin polymerization, as previously noted (crawling). In addition, cancer cells secrete tissue-dissolving substances called *metalloproteinases*, which help clear a pathway for their transport.

Because many cells are integral parts of tissues, they are exposed to numerous mechanical stresses as part of the organisms' anatomy and physiology. The so-called cytoskeleton is the protein-based structural support found in eukaryotic cells, analogous to the bony skeleton of our bodies. The cytoskeleton components include intermediate filaments, microtubules, and actin fibers. These components vary in their elasticity much in the same way that bone and cartilage vary to provide a range of support and flexibility. In addition, microtubules function with so-called motor proteins to move cell organelles, such as Golgi apparatus, mitochondria, cell nucleus, and so on, to judiciously located positions within the cell. Cellular anatomy is a critical factor in stem cell behavior and differentiation, notably in embryonic development. We note that microtubules also can function in the transport pathway system previously described to help rapidly move various species within the cell.[10]

---

[9] For more details, see Alberts et al. (2002).

[10] See, e.g., Alberts et al. (2002).

Information Systems, Check Pointing/Safety, and Reliability

Genetic material is the primary information storage and resource for the living cell. A human cell nucleus contains 46 total chromosomes, of which 23 are paternal and 23 are maternal, or 23 pairs of chromosomes. Each chromosome is a supercoiled strand of DNA consisting of the Watson-Crick base pairing introduced in Chapter 1. There are approximately 2.9×10^9 total base pairs in the human genome, and approximately 30,000 protein coding regions, or genes, have been identified. An example of the chromosomal location of a particular gene, the CFTR gene associated with the disease cystic fibrosis, is illustrated in Figure 2-8.

Unfortunately, cellular processes are not error-free, and mistakes are made continually. For example, in the DNA synthesis

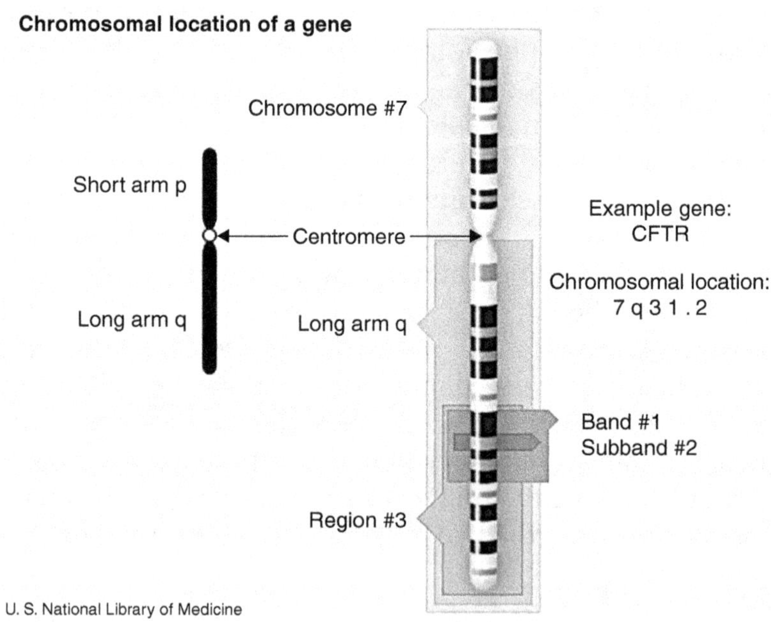

FIGURE 2-8 The standard numerical calling sequence for the location of a particular gene on a chromosome. In this case the human CFTR gene (7q31.2) on chromosome 7. Reproduced from the National Library of Medicine, by permission.

phase of mitosis (copying of DNA prior to division), one incorrect base pairing is made for about every 10^7 bases copied. For example, a common culprit is a uracil residue being mistakenly placed for a thymine residue. Many times an incorrectly copied base pair will have no effect on a daughter cell if it is not involved in the coding region. However, just one incorrectly copied base pair (a single nucleotide polymorphism, or SNP) in the coding region can have disastrous consequences to the entire organism. For example, diseases such as sickle cell anemia, Alzheimer's, cystic fibrosis, cancer, muscular dystrophy, and many others are the result of only one SNP in a particular gene. Because mistakes are so frequent, the cell has special machinery to try to police its many tasks. In the case of DNA synthesis, DNA polymerases contain proof-reading enzymes as part of their total structure, and mistakes are corrected during the synthesis process. This mismatch repair mechanism decreases the error rate to about 1 in 10^9 bases copied, and the cell has additional mechanisms to handle these relatively infrequent mistakes and others.

Cellular DNA is continuously damaged by external factors, such as by ionizing radiation, UV light, and various types of chemicals. DNA damage within a cell that is not repaired results in a mutation. The cell has specific sensor proteins that continuously scan its DNA strands for errors or damage. Once DNA damage has been detected, the cell invokes one or more of a number of different processes, including repair via unique DNA repair enzymes, apoptosis, or programmed cell death, and others as shown in Figure 2-9. It is interesting that genetic defects associated with the making of repair enzymes themselves can lead to disease. For example, xeroderma pigmentosum is a disease whereby a repair enzyme (protein) is either not made or made incorrectly. Individuals with this disease are prone to DNA defects or mutations.

In addition to defects in DNA, defects in the so-called folding process during protein synthesis (translation) also are common. As new proteins are synthesized, they fold into their final active conformational state, which typically is a hydrophobic

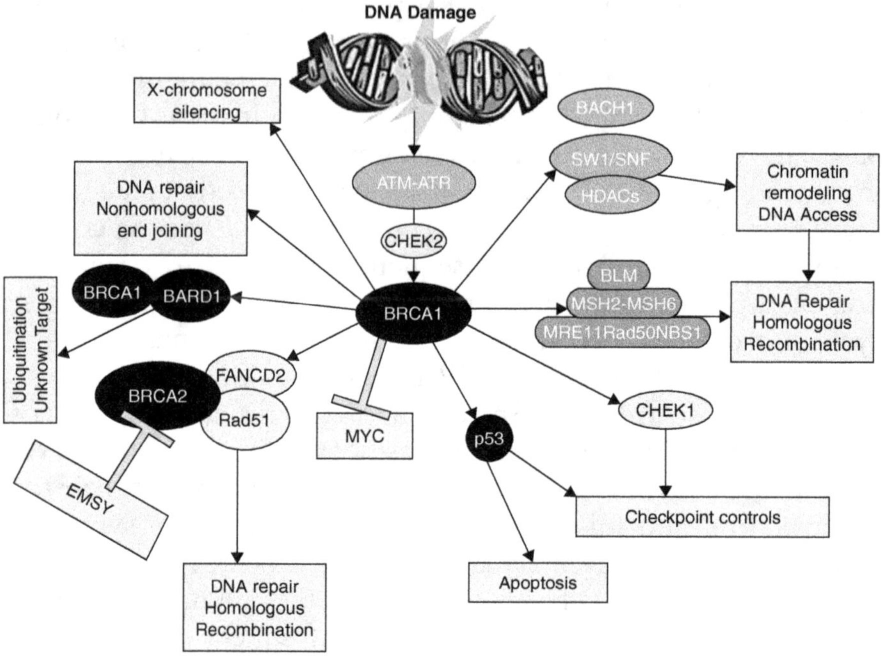

FIGURE 2-9 DNA damage and repair example. Illustration of mutations in the tumor suppressor genes BRCA1 and BRCA2, which predispose women to breast and ovarian cancer. From Dr. B. Friedenson, Medscape General Medicine, 2005;7(2):60, ©2005, by permission.

interior core and a hydrophilic protein exterior. Cells contain chaperone proteins that help produce the correct folded state of proteins, since different folded states may have similar thermodynamic free-energy minimums. When mistakes are made, proteases within the cell break down or destroy unfolded or poorly folded proteins. As with DNA, when unfolded proteins are allowed to persist in the cell, disease can be the unfortunate result. Amyloid deposits—for example, in Alzheimer's disease—are the result of poorly folded proteins that tend to aggregate, forming insoluble fibers.

There is important terminology with regard to the information system of a cell and with the characterization of cell type. *Genotyping* is a term used to describe the characterization of an organism on the basis of its DNA. Genotyping involves the study of genes and their variants (alleles) that help characterize the organism (hair color, blood type, etc.). The *phenotype,* on the other hand, is the physical and chemical characterization of a cell or organism. For example, a phenotypical characterization of islet cells in the pancreas is the production of the endocrine cell signaling protein insulin. In this case, the phenotype can be traced to a particular gene—but in other cases, the phenotype may be associated with interactions of the organism with the environment and may not be of genetic origin. For example, changes in the nutrient levels or temperature and pH of the cellular surroundings may change a cell phenotype, but not its genotype.

Numerous cellular proteins are responsible for the policing of cellular processes and functions. Breakdowns or inefficiencies in policing can lead to changes in genotype (mutations) and phenotype (expressions/characterizations), and this can have disastrous consequences on cell behavior as it relates to the organism as a whole.

Finally, we note that safety is another area of cell function as it relates to the immune system of its parent organism. The so-called adaptive immune system in animals can sense foreign cells and substances through the various antigens that these substances present. Once foreign substances are detected, the organism launches a series of cellular attacks against the foreign invaders (immune response). In order to protect the organism's own cells from such an attack, cells present their own antigens on their surface that the immune system recognizes as self. These antigens are known as *major histocompatibility complexes (MHCs),* and they serve as a flag to keep the immune surveillance system at bay. Breakdowns in this internal cell safety system also can lead to devastating diseases known as *autoimmune diseases,* such as rheumatoid arthritis, multiple sclerosis, diabetes, and many others.[11] In such cases, the organism's immune system attacks its own cells. This is also the mechanism whereby allologous transplant tissue is rejected by the recipient organism.

---

[11] See, for example, Abbas and Lichtman (2001).

Time and Length Scales

Quantification of the cellular processes just discussed begins with an examination of the length and time scales of the underlying phenomena. As we will see in Chapter 3, length and time scales dictate the particular approach to quantifying cellular events. For example, the process of solvating an ion takes place on time scales on the order of picoseconds and on length scales on the order of nanometers. Quantification of this process requires a purely molecular level dynamic analysis. The process of a paracrine signal molecule transport takes place on time scales of milliseconds and length scales of microns to millimeters. In this latter case, so-called transport theory approaches are necessary. One of the key concepts in Chapter 3 is to illustrate how these seemingly widely different analyses are intimately connected.

Figures 2-10 and 2-11 give some typical examples of cellular length scales and time scales, respectively. The subcellular

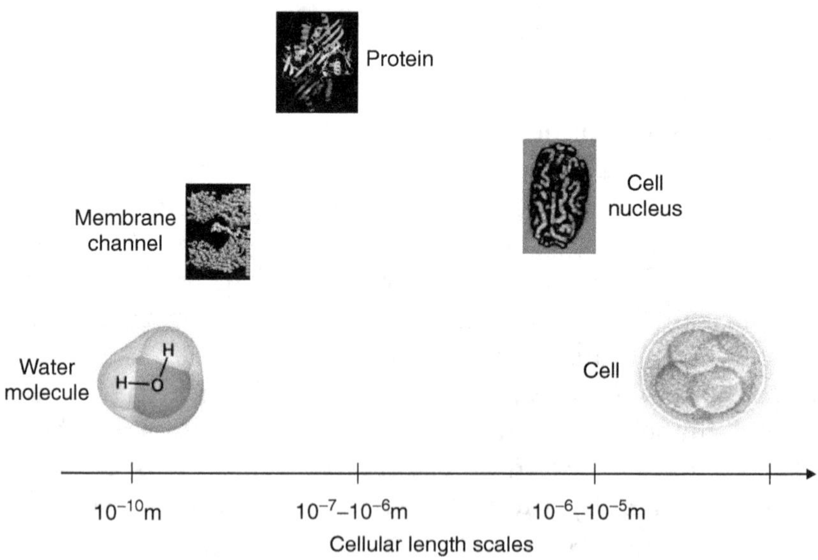

FIGURE 2-10 Illustration of the wide range of length scales in cellular systems. The overall goal in an engineering analysis is to seamlessly connect both length and time scales to provide a complete dynamic description of the living cell.

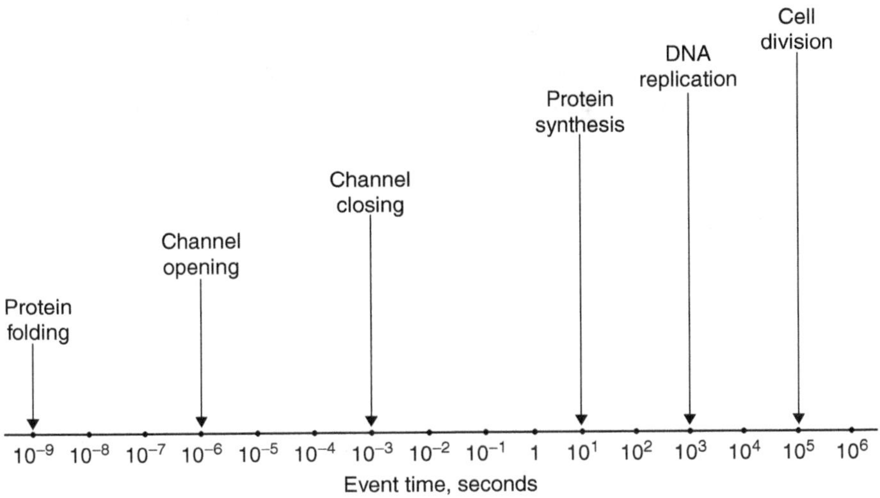

FIGURE 2-11 A companion illustration to Figure 2-10 for various event time scales in cellular systems. The time disparities are equal to or greater than the length disparities.

systems previously described interact in highly orchestrated ways, so we *must* include time (dynamics) as an explicit variable. In the overall quantification of cellular processes and dynamics, these disparate length and time scales must somehow be tied together. As mentioned in the description of biochemical production systems, mathematical nonlinearities often play a critical role in helping to orchestrate these events. These nonlinearities are associated with an accurate physicochemical description of the system or process, and that is where we begin in the next chapter.

References

Abbas, A.K., and Lichtman, A.H., *Basic Immunology,* W.B. Saunders, New York (2001).

Alberts, B., Bray, D., Hopkin, K., Johnson, A., Lewis, J., Raff, M., Roberts, K., and Walter, P., *Molecular Biology of the Cell,* 4th ed., Garland, New York (2002).

Bradshaw, R.A., and Dennis, E.A., *Handbook of Cell Signaling,* Academic Press, New York (2003).

Gilbert, C.S.F., *Developmental Biology,* 7th ed., Sinauer, MA 2003.

Gomperts, B.D., *Signal Transduction,* Academic Press, New York (2003).

Heinrich, R., and Schuster, S., *The Regulation of Cellular Systems,* Chapman & Hall, New York (1996).

Hille, B., *Ion Channels of Excitable Membranes,* Sinauer, MA (2001).

Kiessling A.A., and Anderson, S., *Human Embryonic Stem Cells,* Jones and Bartlett, Boston (2003).

Voet, D., Voet, J.G., and Pratt, C.W., *Fundamentals of Biochemistry,* 2d ed., Wiley, New York (2006).

Chapter Three

The Dynamics of Single, Isolated Biological Macromolecules

As discussed in Chapters 1 and 2, molecular sizes are typically on the order of angstroms (10^{-10} *meters*), and molecules move over spatial scales on this order in femtoseconds (10^{-15} *seconds*) of real time. On the other hand, biological macromolecules have relatively large sizes, must travel distances in cellular systems on the order of microns (10^{-6} *meters*) to millimeters, and act over time scales of milliseconds to seconds. Attempting to computationally predict biological macromolecular behavior over the relatively large spatial scales and long times that are needed to describe their real time and space behavior is well beyond the reach of even the very largest of computers. Then, considering the need to describe the tens of thousands of interactions of those macromolecules needed to predict overall cellular behavior may lead one to follow Ludwig Boltzmann over the Bay of Duino. Analytical mathematics, however, can be a powerful tool to circumvent many of these seemingly hopeless computational obstacles. In particular, applied mathematicians have shown us that complete mathematical descriptions of physical problems seldom are needed within certain measurable degrees of approximation. In other words, there are significant parts of the mathematical description that may play a more or less dominant role in any given system. Identification of mathematically dominant terms in physical descriptions is therefore a potentially powerful tool in making headway on predicting the real-time behavior of biological macromolecules and their interactions.

Dominant Term Analysis

What is really quite satisfying about the tools of applied mathematics and dominant term analysis is that like all mathematics they lead to particular unyielding results whether we like them or not. In other words, sometimes simplifications may result and sometimes they may not. Fortunately, many problems of nature exhibit some kind of term or terms dominance. Dominant term analysis begins with so-called variable scaling, which is discussed next.

In general, scaling involves the nondimensionalization of both the dependent and independent problem variables, where the characteristic variables used for nondimensionalization are selected as their *maximum* values—for example, the maximum values of velocity, temperature, length, and so on, in a particular problem. Let's see specifically how this method works and why it can often lead to mathematical simplifications.

Consider, for example, temperature (T) and velocity ($\mathbf{v}$) as dependent variables and position ($\mathbf{r}$) and time (t) as independent variables, and we introduce dimensionless forms by an asterisk as

$$T^* = T/T_0$$
$$\mathbf{v}^* = \mathbf{v}/v_0$$
$$\mathbf{r}^* = \mathbf{r}/r_0$$
$$t^* = t/t_0$$

where the characteristic variables T_0, v_0, r_0, t_0, used for nondimensionalization are to be selected in a "special" manner. Specifically, in scaling the characteristic variables are selected as the maximum values that exist in that particular problem. In doing this, the scaled variables are confined to the domain [0,1], and we are attempting to set the *rate of change* of the dependent variables close to 1 throughout the domain in question. This underlying principle is illustrated in Figure 3-1 for temperature as a function of the spatial variable x. As seen in Figure 3-1, both the scaled temperature and distance vary between 0 and 1, and the rate of change of temperature with distance is approximately unity throughout the domain. The behavior shown in

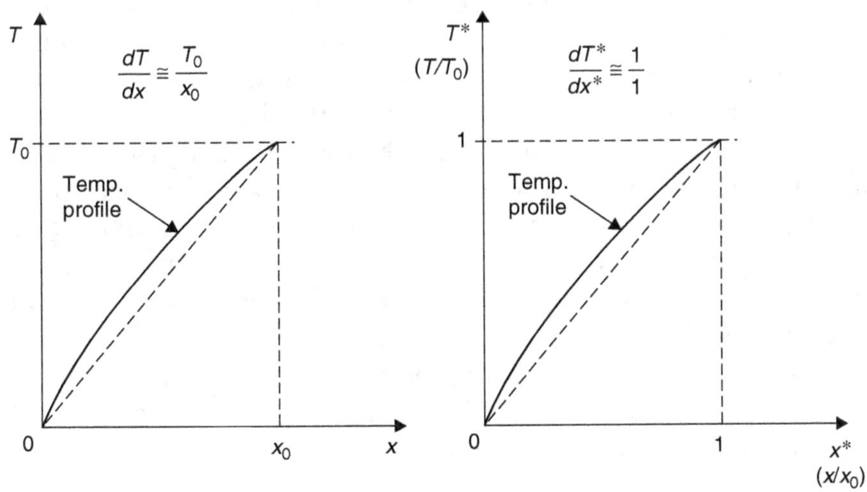

FIGURE 3-1 Orthodox behavior of temperature as a function of distance. The dimensionless spatial derivative is on the order of unity throughout the domain.

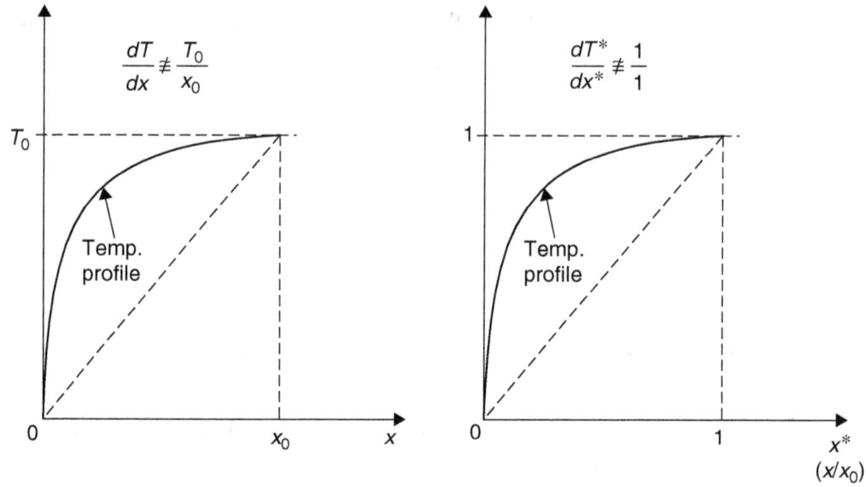

FIGURE 3-2 Unorthodox behavior of temperature as a function of distance. The dimensionless spatial derivative deviates from unity in at least some parts of the domain. Multiple scaling parameters are required in this case.

Figure 3-1 is called *orthodox,* meaning "commonly observed." However, if the dependent variables (velocity or temperature) vary dramatically over the domain (Figure 3-2), then the dimensionless derivatives will not necessarily be close to 1 throughout the domain. In this latter case, the behavior of the dependent variable is said to be *unorthodox* with respect to the independent variable, and using the maximum values for the characteristic variables does not properly scale the derivative term throughout the entire domain in question.[1] In this latter case, it often is possible to use different scales or characteristic variables in different parts of the domain. These so-called multiple-scale methods are quite powerful, and we will see a particular application here in predicting the long-time dynamics of biological macromolecules.

As a simple example of orthodox behavior and its significance in mathematical simplification, consider the problem of two-dimensional energy conduction through a "long" solid bar

[1] For further examples and discussion, see Lin and Segel (1974).

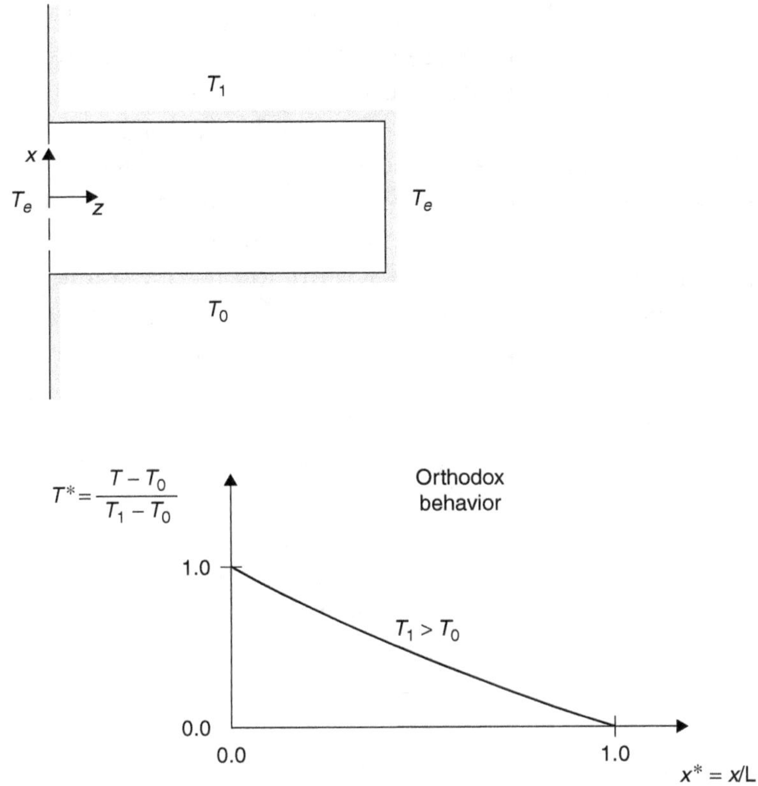

FIGURE 3-3 Temperature profiles in a long solid bar. For long thin bars, scaling demonstrates that conduction along the length of the bar may be neglected.

as shown in Figure 3-3. The temperatures along the boundaries are maintained as shown in Figure 3-3. With the dimensionless, scaled temperatures defined in this particular case as $T^* = (T - T_0)/(T_1 - T_0)$, and spatial variables defined as $x^* = x/x_0$ and $z^* = z/z_0$, the steady-state energy conduction equation follows the dimensionless Laplace form as

$$\frac{\partial^2 T^*}{\partial x^{*2}} + \left(\frac{x_0}{z_0}\right)^2 \frac{\partial^2 T^*}{\partial z^{*2}} = 0$$

Our particular scaling has resulted in both $\partial^2 T^*/\partial x^{*2}$ and $\partial^2 T^*/\partial z^{*2}$ having approximate values of 1 or less, except at the ends (Figure 3-3). Now with $z_0 >> x_0$ (long bar), the coefficient of the second term (z-conduction term) will always be extremely small. Therefore, we can safely neglect the second term compared to the first term—that is, as long as we stay away from the very ends of the bar. This is the mathematical reasoning behind neglecting conduction along the length of a bar for extremely long bars and, of course, it greatly simplifies the mathematical problem by reducing it to a single spatial dimension. Note that this results in a very simple ordinary differential equation to solve as opposed to the relatively more difficult Laplace's partial differential equation.

As a common example of unorthodox behavior, consider the large velocity (large Reynolds number) flow over a sphere as shown in Figure 3-4. We might initially assume that the characteristic length scale in this problem is the sphere diameter (d)—that is, the well-known Reynolds number, which is the ratio of fluid inertial to viscous terms, is postulated to be

$$N_{\mathrm{Re}} = \frac{\rho d v_\infty}{\mu}$$

where ρ and μ are the density and viscosity of the fluid, respectively, and v_∞ is the so-called free stream fluid velocity. Thus,

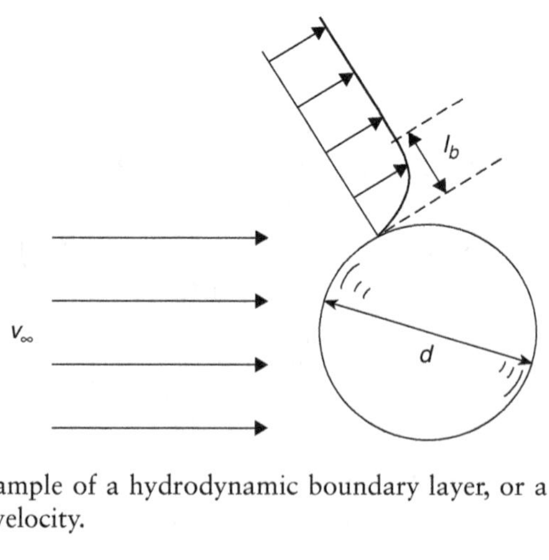

FIGURE 3-4 Example of a hydrodynamic boundary layer, or a layer of dramatic variation in fluid velocity.

we are led to the condition $N_{\text{Re}} >> 1$ for large v_∞ and that the viscous terms should be negligible under this condition. The actual velocity field is shown in Figure 3-4 and, as can be seen, the velocity changes sharply over a very thin region near the surface known as the *momentum boundary layer region*. The proper length scale in this region is the thickness of the boundary layer (l_b), which is known to be much, much less than d. The proper Reynolds number in the boundary layer region is actually

$$N_{\text{Re}} = \frac{\rho l_b v_\infty}{\mu}$$

which is typically *not* large even for large v_∞. In this problem, we can assign different length scales in different parts of the domain (near sphere and far sphere) and the solutions can be formally, asymptotically matched together in a sort of a mathematical patching or mending exercise.[2] The techniques fall under the general subject of perturbation theory, and we will have occasion to refer to them here as they play a critical role in the analytical mathematical treatment of dynamic processes in biological systems.

The particular variables in biological macromolecular systems notably include time, the spatial variables associated with the particular molecular structure and surroundings, and the forces of interactions of biological macromolecules. These forces include intramolecular forces between the atoms within any particular biological macromolecule, interaction forces between two or more biological macromolecules, and the forces between biological macromolecules and the surrounding solvent. Interaction forces are often given in terms of a potential function, so it is important to understand the relationship between them, in addition to the types of forces, as discussed here.

---

[2] See Kervorkian and Cole (1981), Van Dyke (1975), and Nayfeh (1973).

Force Descriptions

In general, elementary intermolecular or interatomic interaction forces can be classified as follows:

- Electrostatic forces
- van der Waals forces
- Born repulsive forces
- covalent bonding (spring forces, angle forces, etc.)

Coulombic electrostatic interaction forces exist between charged molecules or atoms. The molecular or atomic charging results from the condition of electron deficiency or electron excess in the atoms. Other common electrostatic forces are dipole-dipole interactions among polar molecules, such as water, and higher-order multipole interactions. Van der Waals forces are associated with oscillating, time-dependent electric fields created by the wave, or probabilistic behavior, of electrons in atoms. The van der Waals forces between atoms are typically attractive in nature and vary as r^{-7}, where r is the interatomic separation distance. Born repulsive forces result when two atoms or molecules approach extremely close to one another at distances slightly larger than σ, where σ is the atomic (molecular) diameter. At these very small separation distances, the electron shells of two interacting atoms or molecules can partly penetrate one another. The nuclear charge interatomic interactions, however, which no longer are fully screened by the electrons, cause strong repulsive forces between the atoms or molecules, preventing further overlap. Repulsive forces are short-range forces compared to van der Waals forces. Covalent bonding results when an electron becomes shared between two adjoining atoms, and it is included here as an interaction force because it often is classically modeled in such a manner, as discussed more fully here.

Consider the force (type 1, 2, or 3 above) acting on atom or molecule 1 by atom or molecule 2, as shown in Figure 3-5. For *central forces,* we have

$$\mathbf{F} = F_r \mathbf{e}_r$$

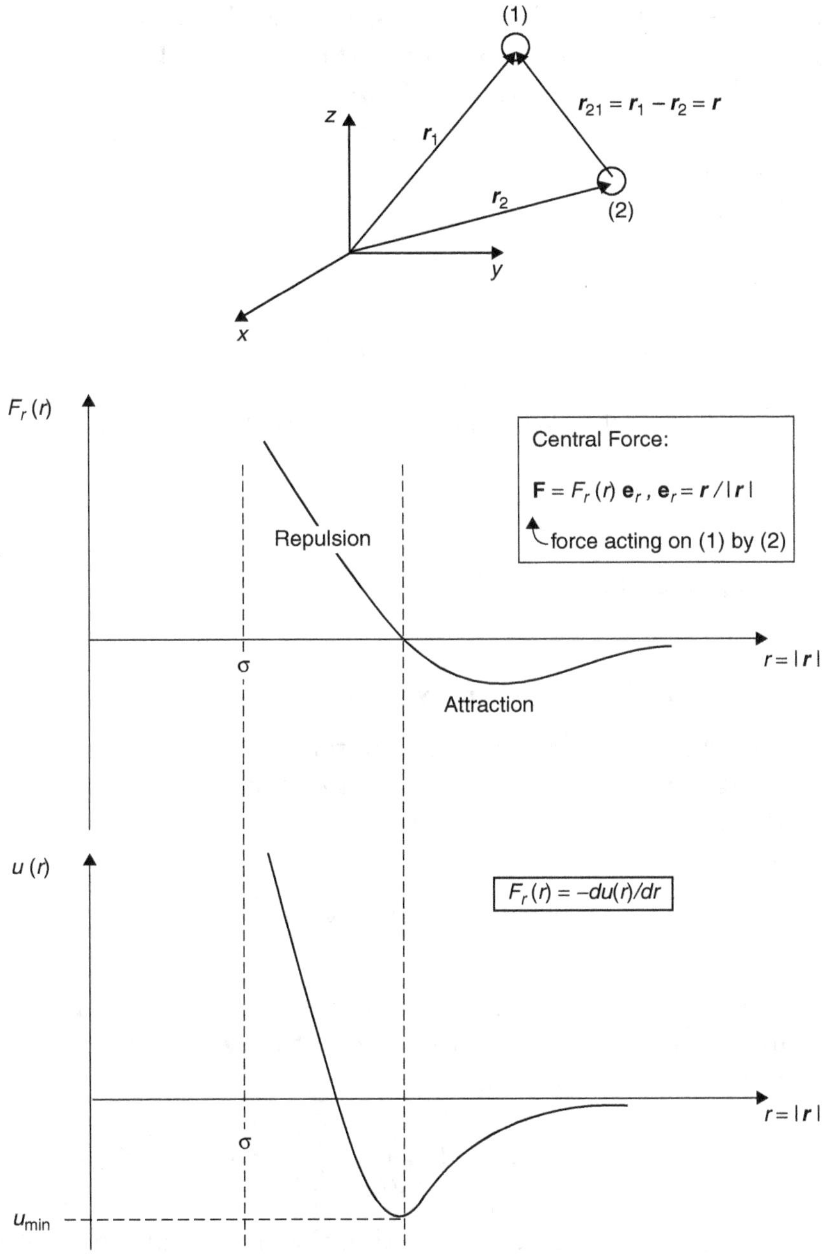

FIGURE 3-5 Typical interatomic force or potential interaction profile, such as a Lennard-Jones force or potential field. Born repulsive forces dominate at close separations preventing atomic overlap, whereas attractive van der Waals forces dominate at larger separation distances.

where $\mathbf{e}_r = \mathbf{r}/|\mathbf{r}| = \mathbf{r}/r$ is a unit vector pointing from atom or molecule 2 to atom or molecule 1 (Figure 3-5). If $F_r < 0$, the force acting on 1 is attractive, and if $F_r > 0$, the force is repulsive, as shown in Figure 3-5. The potential of the force is defined according to

$$F_r(r) = -du(r)/dr$$

Integrating

$$-\int_{\infty}^{r} F_r(r')dr' = \int_{\infty}^{r} du = u(r) - u(\infty) = u(r)$$

where we have arbitrarily set $u(\infty) = 0$ as the reference potential. Thus,

$$u(r) = \int_{r}^{\infty} F_r(r')dr'$$

This is sketched in Figure 3-5.

Modeling Born Repulsive and van der Waals Forces

Intermolecular (or interatomic) potentials characterized by short-range (Born) repulsive forces and longer-range (van der Waals) attractive forces often are mathematically modeled using the so-called Mie potential expression

$$u(r) = -\frac{A}{r^m} + \frac{B}{r^n}$$

where the first term accounts for attraction, and the second term accounts for repulsion. A useful form of the Mie potential is the Lennard-Jones 6-12 potential

$$u(r) = 4\epsilon\left[-\left(\frac{\sigma}{r}\right)^6 + \left(\frac{\sigma}{r}\right)^{12}\right]$$

where ϵ and σ are constants related to the energy minimum and distance at minimum, respectively (Figure 3-5).

The Lennard-Jones (LJ) model is a good representation for intermolecular interactions involving electrically neutral, nonpolar, and short/compact-shaped molecules. A short list of the

Table 3-1 Lennard-Jones 6-12 potential constants*

Substance	ϵ (kcal/mol)	σ(Å)
N—N	0.170	3.250
O—O	0.210	2.960
C—C	0.086	3.400
S—S	0.250	3.564
P—P	0.200	3.742
OW—OW	0.152	3.151
CT—CT	0.109	2.960
OH—OH	0.210	3.066

*AMBER Force Field values; Cornell et al. (1995).
Notes: C—C includes all carbon types except aliphatic carbon (CT); N—N includes all nitrogen types; O—O includes all oxygen types except water (OW) and hydroxyl oxygen (OH).

popular Lennard-Jones 6-12 potential constants for biological atoms is given in Table 3-1. These values are the self atomic interactions, such as N—N, O—O, and so on, and so-called sum rules are used to determine non-self interactions as[3]

$$\epsilon_{ij} = (\epsilon_i \epsilon_j)^{1/2}$$

$$\sigma_{ij} = (\sigma_i + \sigma_j)/2$$

where i and j are any two interacting atoms.

Models for Electrostatic Interactions

Many biological molecules possess charges, due to electron deficiency or excess of their constituent atoms, as well as dipole, quadrapole, or higher moments, due to shifting of electron positions (in a time-averaged sense) between bonded atoms. The LJ potential model must be supplemented to describe electrostatic intermolecular interactions in these systems.

The simplest electrostatic interaction is the Coulomb potential between two point charges, q_1, q_2, located a distance r apart:

$$u(r) = \frac{q_1 q_2}{r}$$

where q_1 and q_2 are given in electrostatic units (1 esu $\equiv$ 1 dyne$^{1/2}\cdot$ cm). The Coulomb potential, along with a Born

[3] Hirschfelder et al. (1949).

repulsive potential, often is a good model for simple ionic interactions, such as interactions between Na^+ and Cl^-. When the intervening matter between the two interacting point charges is electrically nonneutral, such as water (a highly polar solvent), it is necessary to augment Coulomb's law, for example, through the use of a dielectric constant in the previous electrostatic force expression. These particular issues are treated more fully in Chapter 4 when interacting biological macromolecules are considered.

For molecules possessing dipole, or higher, moments and for molecules with localized charge groups, such as proteins and nucleic acids, the subatomic structure of the molecule must be taken into account in the electrostatic description that necessitates a quantum mechanical solution. A common classical approach in modeling the electrostatic interactions in these systems is to assign partial point charges to the atoms comprising the molecule that produce the proper overall charge effect. An important illustration of this is the simple point charge model for water, or SPC model discussed here.

As noted in Chapter 1, water is a highly polar molecule with oxygen being much more electronegative than hydrogen. This molecular system can be effectively modeled electrostatically by assigning partial point charges to each of the oxygen and hydrogen atoms, as shown in Figure 3-6.[4] The partial point charges have been shown to create the correct dipole moment and other physical properties of bulk water. In addition, the covalent O—H bonds in water exhibit distance flexibility similar to spring connectors. Spring connector forces are often used to represent covalent bonding that allows for a certain degree of internal molecule flexibility. Typically, the springs are linear with constants obtained from simple potential field expansions about an equilibrium state (small oscillation theory), as shown in the next section. In the case of water, the equilibrium distances are given in Figure 3-6, and the small oscillation theory results for the spring constants are given by Toukau and Rahman (1985)

[4] Toukau and Rahman (1985).

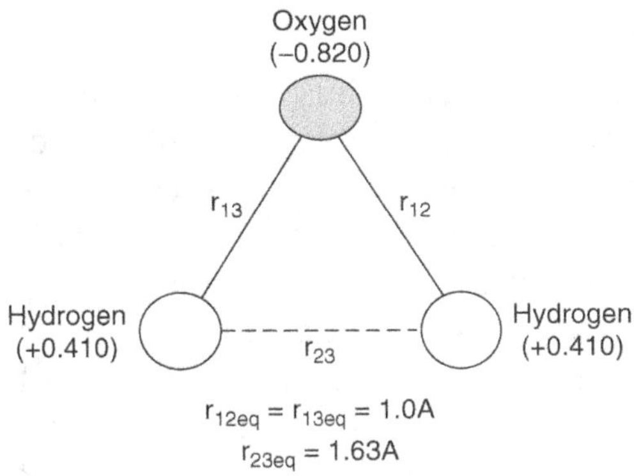

FIGURE 3-6 The SPC model for a water molecule. Atomic charges centered at the atom center of mass are indicated in parentheses. Vibrations result in variations in the interatomic separation distances about their equilibrium values, as shown.

and are discussed in more detail in the section on covalent bonding.

Locally charged and polar groups within biological molecules and macromolecules also can be assigned partial point charges at atomic locations. For example, the particular partial point charges for the amino acids comprising peptides and proteins are given in Table 3-2 based on the work of Cornell and others (1995).

Table 3-2 Example of partial atomic charge values in peptides and proteins**

Amino Acid Residue Name	Atom Name	Charge (esu)
Alanine (ALA)	N	−0.4157
ALA	HN	0.2719
ALA	CA	0.0337
ALA	HA	0.0823
ALA	CB	−0.1825
ALA	HB1, 2, 3	0.0603
ALA	C	0.5973
ALA	O	−0.5679
Glycine (GLY)	N	−0.4157
GLY	HN	0.2719
GLY	CA	−0.0252
GLY	HA1, 2	0.0698
GLY	C	0.5973
GLY	O	−0.5679

**Source: AMBER Force Field values: Cornell et al. (1995). Atom naming here conforms to the Protein Data Bank: Berman et al. (2000).

Covalent Bonding

Covalent bonding between atoms of biological molecules varies from being fairly rigid to exhibiting some degree of flexibility. Internal flexibility of biological molecules and macromolecules can be effectively described using small oscillation theory, but the number of possible connections can become large, and the integration with the overall macromolecular dynamics is less developed in the literature. An important example of the use of small oscillation theory is in the modeling of covalent bonding (O—H) or vibrational modes in water molecules. In general, small oscillation theory is based on a multiterm, quadratic-order Taylor series expansion of the total potential energy of interaction between each of the atomic constituents of the molecule.

$$U'(\chi_1, \chi_2, \ldots) = U'(a, b, \ldots) + (\chi_1 - a)\frac{\partial U'}{\partial \chi_1}|_{\chi_1 = a, \chi_2 = b, \ldots} + \cdots + (\chi_1 - a)^2 \frac{\partial^2 U'}{\partial \chi_1^2}|_{\chi_1 = a, \chi_2 = b, \ldots} + \cdots$$

where $\chi_1, \chi_2, \ldots$ are the independent coordinates and $a, b, \ldots$ are their constant, equilibrium values, respectively.

Identification of the particular degrees of freedom and independent coordinates is a necessary first step to expand the potential about its equilibrium structure. The linear force acting on any atom in the lab frame can then be obtained by differentiation of the quadratic potential, where the independent coordinates must be related to the laboratory coordinate system, for example, the force acting on atom 1 is given by

$$\mathbf{F}_1 = -\frac{\partial U}{\partial \mathbf{r}_1}$$

where $\mathbf{r}_1$ is a laboratory frame locator vector for atom 1, and U must be related and expressed in terms of the laboratory frame coordinates:

$$U'(\chi_1, \chi_2, \ldots) = U(\mathbf{r}_1, \mathbf{r}_2, \ldots)$$

where $\mathbf{r}_1, \mathbf{r}_2, \ldots$ are the laboratory frame locator coordinates for atom 1, atom 2, and so on. Torques also may exist based on changes in the potential with respect to the Euler angles.[5] For those not familiar with Euler angle descriptions, a brief tutorial is given in Appendix A.

Following this analysis, specific expressions for internal water molecular vibrations have been obtained by Toukau and Rahman (1985). Internal vibrations and rotations in biological macromolecules also are treatable in a similar manner. For now, however, we will concentrate on multiple-scale dynamic descriptions for *rigid* macromolecules in water.

Molecular Dynamics and Brownian Dynamics for Isolated Biological Macromolecules

In general, classical molecular dynamic simulations of biological macromolecules first involve specifying the macromolecular structure, such as through the popular Protein Data Bank (*.pdb files),[6] and then hydrating or placing the macromolecule in the appropriate molecularly specified solvent. Hydration schemes are numerous in the literature and include grand canonical Monte Carlo methods,[7] SOLVATE,[8] and others. Each atom of the system (macromolecule and solvent atoms) then is moved according to finite difference representations of Newton's laws of motion.[9] This requires very small time steps on the order of femtoseconds for each of the finite differential atomic movements. In order to properly simulate the effect of the solvent on the macromolecular motion requires tens of thousands or more solvent molecules in a relatively large simulation box called the *control volume*. So-called periodic boundary conditions are used along the surface of the control volume to simulate an

[5] Peters and Ying (1993).

[6] Berman et al. (2000).

[7] See, e.g., Heermann (1990), and Allen and Tildesley (1989).

[8] SOLVATE is written by Helmut Grubmmller, Theoretical Biophysics Group, Institut für Medizinische Optik, Ludwig-Maximilians-Universität München, München, Germany.

[9] For a more comprehensive description of molecular dynamics methods, see Heermann (1990), and Allen and Tildesley (1989).

infinite solvent system; however, other boundaries can also be specified. For constant temperature simulations or canonical ensembles, periodic system temperature adjustments, based on the net relative atomic kinetic energies, are necessary. Finally, the dipole nature of water and any solvent ions require special treatment of their interactions due to the relatively long-range nature of their associated electrostatic forces.[10]

As it turns out, it is not necessary to actually conduct such brute force molecular dynamic simulations of biological macromolecules over the entire time scale of their movement. A little bit of applied mathematics and physics demonstrates that there are at least two disparate time scales involved in the overall motion that simplifies the problem considerably:

1. For any given macromolecular position and orientation, the surrounding solvent molecules quickly relax to a local equilibrium state in the potential field created by the macromolecular atoms (i.e., the field created by their charge, Lennard-Jones potential, etc.). This solvent relaxation takes place on a time scale of picoseconds (10^{-12} seconds) or less.
2. No appreciable change in the macromolecular position or orientation occurs during the solvent relaxation period, and small, appreciable differential changes for rigid macromolecules (overall translation and rotational motions) take place on time scales on the order of microseconds (10^{-6} seconds) to nanoseconds (10^{-9} seconds), depending on the macromolecular size and the strength of any external field forces.

Thus, there is potentially a million or more orders of magnitude difference in these two particular time scales!

As noted earlier in this chapter, multiple-scale methods can be used to efficiently simplify this system and avoid inclusion of nondominant regions. In this particular case, two time scales

[10] A useful and effective approach is the so-called the reaction field method; see Tironi et al. (1995).

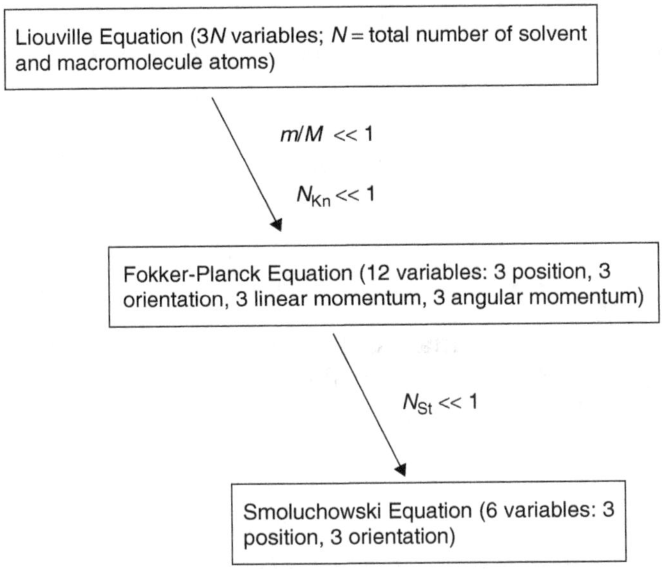

FIGURE 3-7 Schematic diagram of the steps in development of the multiple time scale algorithm. Mathematical details are given in Appendix B.

can be formally introduced into the problem, and the two associated solutions: a short-time one describing the fluid relaxation problem and a long-time one describing the bulk macromolecular motion (called *Brownian motion*, or *Brownian dynamics*). These two solutions are formally asymptotically matched together to provide a complete spatial and temporal description of the problem.[11]

Appendix B reviews the essential mathematical elements of the overall multiple-scale analysis and provides key results. A pictorial summary of the analysis is given in Figure 3-7. One begins with the Liouville equation, which is the fundamental starting equation describing the complete molecular system dynamics for all times. It can be demonstrated that particular mass and length scale expansions lead to the simpler Fokker-Planck

[11] Cukier and Deutch (1969). Inclusion of rotational and rotational-translational coupling motions were given using so-called time-smoothing methods by Condiff and Dahler (1966) and by formal, multiple time scale methods, including external surface interactions, by Peters (1999, 2000).

(FP) equation for a biological macromolecule in a host solvent. Specifically, the FP equation holds under the conditions

$$\frac{m}{M} << 1$$

and

$$N_{\mathrm{Kn}} << 1$$

where m is the average mass of a molecule of the solvent and M is the total mass of the macromolecule, and N_{Kn} is the so-called Knudsen number defined as

$$N_{\mathrm{Kn}} = \frac{r_0}{R_0}$$

In this last definition, r_0 is a characteristic length scale for the fluid intermolecular interaction forces and, for an isolated Brownian particle, R_0 is a characteristic length scale for the interaction force between the macromolecule and the fluid, often taken as the effective overall macromolecular diameter. Both the mass ratio and length scale ratios must be small for the FP equation to hold. Physically, the mass ratio requirement means that the macromolecule will not be moved or altered by an interaction with a single solvent molecule; it will take many such interactions to effect appreciable changes. The length scale ratio is necessary if the solvent is to quickly relax to a local solvent equilibrium state in the presence of the macromolecule. It is possible to extend the FP analysis to other situations, such as particles in gases, but this is not relevant to our situation here.

The development of the FP equation leads directly to the two time-scale calculation method. The short-time behavior, typically over picoseconds, is a molecular dynamics method that describes the solvent relaxation in the presence of the macromolecule without any appreciable change in the macromolecular position, orientation, or linear and angular momenta ($3N$ independent variables, where N is equal to the total number of solvent and macromolecule atoms). The solvent relaxation behavior in the presence of the macromolecule establishes the parameters, such as its mobility, net forces, and so on, necessary to move the macromolecule over a long-time scale (Brownian dynamics step). The long-time behavior, typically over nanoseconds to microseconds, uniquely describes the subsequent change of the macromolecule's position, orientation, and linear and angular momenta (12 independent variables). After moving the macromolecule, the entire procedure is repeated—that is, resolvation and solvent relaxation followed by macromolecule displacement.

Finally, if valid, a particular type of force scale term expansion leads from the FP equation to the even simpler Smoluchowski (Sm) equation to describe the long-time behavior. Specifically, for the Sm equation to hold, we must have

$$N_{\mathrm{St}} << 1$$

where N_{St} is the macromolecule or particle Stokes number defined as

$$N_{\mathrm{St}} = \frac{\zeta_0^{-1} M}{R_0/(kT/M)^{1/2}} = \frac{t_{\mathrm{relax}}}{t_0}$$

where ζ_0 is a characteristic particle friction coefficient, k is Boltzmann's constant, and T is absolute temperature. Physically, the Stokes number is the ratio of the particle relaxation time $\zeta_0^{-1} M$ to the characteristic time constant for particle displacement t_0.

The Stokes number requirement for validity of the Sm equation can be rewritten as

$$R_0 >> \left(\frac{kT}{M\beta^2}\right)^{1/2}$$

where $\beta \equiv \zeta_0/M$. The last inequality relationship was first given by Chandrasekhar,[12] and it is important in establishing limitations of the Sm equation. For example, in the case of a nonisolated Brownian particle, relatively strong external forces, due to other nearby macromolecules or due to field forces (small characteristic R_0), could negate the applicability of the Sm equation and, thus, requiring the FP equation and FP Brownian dynamics for the long-time description.

[12] Chandrasekhar (1943).

As shown in Figure 3-7, the Sm equation does not require the macromolecular linear and angular momenta as independent variables, and it can be extended to longer time displacements than the FP equation. Both the FP and Sm equations admit numerical solutions that constitute the so-called Brownian dynamics method—that is, either FP or Sm Brownian dynamics, respectively. The mathematical steps are summarized in Appendix B. Here, we present the final useful results for implementing Sm Brownian dynamics (BD) and its coupling to the short-time, molecular dynamics (MD) calculation.

The *long-time* behavior, or Sm BD, describes the total macromolecular motion in space and can be put in the form of *dimensionless* translational and rotational displacement equations describing the change in the position of the center of mass **R** and the orientation about the center of mass ϕ as, respectively,

$$R_i = R_i^0 + \epsilon t \left[\sum_j \left(D^0_{ij_T} F'^0_j + D^0_{ij_{TR}} T'^0_j \right) \right] + C_i$$

$$\phi_i = \phi_i^0 + \epsilon t \left[\sum_j \left(D^0_{ij_{RT}} F'^0_j + D^0_{ij_R} T'^0_j \right) \right] + C_i$$

where superscript 0 indicates that these are the initial specified values—that is, at the beginning of the time step, the subscripts i and j are the laboratory frame coordinate designations (three Cartesian coordinates and three rotations about each coordinate), t is the time step for the displacement, F' and T' are any external forces or torques, respectively, acting on the macromolecule, such as through gravity, an external electric field, or nearby macromolecule or surface, D_T, D_{TR}, D_{RT}, and D_R are the translational-rotational diffusion tensors (each, a 3×3 matrix) that depend on the exact molecular structure of the macromolecule and its interactions with the solvent. They are evaluated at the beginning of the BD time step by MD as shown here. Also, C_i is a multivariate, Gaussian random number with zero mean and variance-covariance given by

$$\langle C_i C_j \rangle = 2 D^0_{ij} t \epsilon$$

D^0_{ij} in this last expression represents a single component of the complete 6×6 grand diffusion tensor. The six total multivariate, random variables, C_i, $(i = 1, 6)$ are generated using standard techniques, such as Cholesky factorization.[13]

Also, in three equations, ϵ is a "smallness" parameter specifically given by

$$\epsilon = \frac{D_0}{R_0(kT/M)^{1/2}} \equiv \frac{1}{N_{\mathrm{Pe}}}$$

where k is Boltzmann's constant, T is the system temperature, $(kT/M)^{1/2}$ is the macromolecular thermal velocity or speed, M is the macromolecular mass, D_0 is a characteristic value for macromolecular diffusion, and $R_0 = kT/F_0$ is a characteristic force length scale, where F_0 is a characteristic external force. N_{Pe} is a type of Peclet number, which is the ratio of the diffusion time scale (R_0^2/D_0) to the characteristic time scale $(t_0 = R_0/(kT/M)^{1/2})$ for particle translation or rotation due to the external force or torque field.

Detailed asymptotic analysis demonstrates that the Peclet number must be large (or ϵ small) for the Smoluchowski equation, or what is called the *diffusion regime,* to hold. For example, strong external force or torque fields can lead to relatively small values of R_0 that may negate the use of the Smoluchowski equation. Another typical violation occurs for relatively large values of M, the particle mass (*inertial effects*). Either of these situations may be remedied by the use of the slightly more cumbersome Fokker-Planck equation that includes the linear and angular momenta of the macromolecule as six additional independent scalar variables.[14]

Now, the diffusion tensors are specifically determined from the short-time, molecular dynamics problem, and for rigid macromolecules in very dilute concentrations (noninteracting macromolecules) they are invariant with respect to time in the *body-frame* of the macromolecule. In other words, they need be determined only once for any given macromolecule.

[13] Brately et al. (1996).

[14] Peters (1999).

The *dimensionless* particle diffusion tensors are given in terms of the *dimensionless* particle friction tensors as[15]

$$\mathbf{D}_T \equiv [\zeta_T - \zeta_{TR} \cdot \zeta_R^{-1} \cdot \zeta_{RT}]^{-1}$$

$$\mathbf{D}_{TR} \equiv [\zeta_{RT} - \zeta_R \cdot \zeta_{TR}^{-1} \cdot \zeta_T]^{-1}$$

$$\mathbf{D}_R \equiv [\zeta_R - \zeta_{RT} \cdot \zeta_T^{-1} \cdot \zeta_{TR}]^{-1}$$

$$\mathbf{D}_{RT} \equiv [\zeta_{TR} - \zeta_T \cdot \zeta_{RT}^{-1} \cdot \zeta_R]^{-1}$$

where the dimensionless translational, rotational, and coupled translational-rotational friction tensors are given by the time autocorrelation relationships determined from the short-time MD step as

$$\zeta_T \equiv \int_0^\infty \left[\langle \mathbf{F}_f(s)\mathbf{F}_f(0)\rangle_{\text{eq}} - \langle \mathbf{F}_f\rangle_{\text{eq}}^2\right] ds$$

$$\zeta_{TR} \equiv \int_0^\infty \left[\langle \mathbf{F}_f(s)\mathbf{T}_f(0)\rangle_{\text{eq}} - \langle \mathbf{F}_f\rangle_{\text{eq}}\langle \mathbf{T}_f\rangle_{\text{eq}}\right] ds$$

$$\zeta_{RT} \equiv \int_0^\infty \left[\langle \mathbf{T}_f(s)\mathbf{F}_f(0)\rangle_{\text{eq}} - \langle \mathbf{T}_f\rangle_{\text{eq}}\langle \mathbf{F}\rangle_{\text{eq}}\right] ds$$

$$\zeta_R \equiv \int_0^\infty \left[\langle \mathbf{T}_f(s)\mathbf{T}_f(0)\rangle_{\text{eq}} - \langle \mathbf{T}_f\rangle_{\text{eq}}^2\right] ds$$

where $\mathbf{F}_f(s)$ and $\mathbf{T}_f(s)$, respectively, are the force and torque acting on the Brownian particle by the fluid at time s. Further, $\langle \mathbf{F}_f\rangle_{\text{eq}}$ and $\langle \mathbf{T}_f\rangle_{\text{eq}}$ are the equilibrium average force and torque, respectively, which are zero for an isolated Brownian particle. In practice, these integrals converge for total times on the order of picoseconds.[16]

Note that for interacting Brownian particles or for a particle interacting with a surface, as in ligand-receptor processes, the equilibrium force and torque are not zero, as is discussed more fully in Chapter 4.

Now, the net force and torque acting on the particle by the fluid are given in terms of the potential energy of interaction

[15] See Peters (2000) or Appendix B for details.
[16] Zhang et al. (2003).

u_{fp} between a solvent atom (i) and macromolecule atom (k), as

$$\mathbf{F}_f = -\frac{\partial}{\partial \mathbf{R}}\left[\sum_i \sum_k u_{fp}\right]$$

$$\mathbf{T}_f = -\frac{\partial}{\partial \Omega}\left[\sum_i \sum_k u_{fp}\right]$$

where Ω represents the set or Euler angles, and $\langle \mathbf{F}_f \rangle_{\text{eq}}$ and $\langle \mathbf{T}_f \rangle_{\text{eq}}$ are the equilibrium average force and torque, respectively, determined through time averaging in the canonical equilibrium ensemble of the molecular dynamics calculation.

Once the initial center of mass position and orientation of the macromolecule are assigned, the new position of the center of mass of the macromolecule and its new orientation with respect to the lab frame are determined from the six explicit scalar equations previously given. The new positions of the individual atoms of the macromolecule are obtained by first computing their position changes from the overall rotational change about the center of mass using a finite rotational transformation formula.[17] In particular, if a_i^0 is the initial i^{th}-component of the position in the lab frame of any atom of the macromolecule *with respect to the macromolecule center of mass,* its new value at time t is given by

$$a_i = \sum_j e_{ij} a_j^0$$

where e_{ij} is a particular component of the finite rotational transformation matrix given in total by

$$e_{ij} = \begin{bmatrix} \cos\Delta\phi_y \cos\Delta\phi_z & -\cos\Delta\phi_y \sin\Delta\phi_z & \sin\Delta\phi_y \\ \begin{array}{c}\sin\Delta\phi_x \sin\Delta\phi_y \cos\Delta\phi_z \\ +\cos\Delta\phi_x \sin\Delta\phi_z\end{array} & \begin{array}{c}-\sin\Delta\phi_x \sin\Delta\phi_y \sin\Delta\phi_z \\ +\cos\Delta\phi_x \cos\Delta\phi_z\end{array} & -\sin\Delta\phi_x \cos\Delta\phi_y \\ \begin{array}{c}-\cos\Delta\phi_x \sin\Delta\phi_y \cos\Delta\phi_z \\ +\sin\Delta\phi_x \sin\Delta\phi_z\end{array} & \begin{array}{c}\cos\Delta\phi_x \sin\Delta\phi_y \sin\Delta\phi_z \\ +\sin\Delta\phi_x \cos\Delta\phi_z\end{array} & \cos\Delta\phi_x \cos\Delta\phi_y \end{bmatrix}$$

[17] Goldstein (1988); also see Peters and Ying (1993).

where

$$\Delta\phi_x = \phi_x - \phi_x^0$$

$$\Delta\phi_y = \phi_y - \phi_y^0$$

$$\Delta\phi_z = \phi_z - \phi_z^0$$

The new positions of each atom of the macromolecule in the lab frame can be obtained as

$$r_i = R_i + a_i$$

To obtain the new diffusion tensor components, the rotational transformations are applied twice according to

$$D_{ij} = \sum_k \sum_l e_{ik} e_{jl} D_{kl}^0$$

Having initially determined the grand diffusion tensor for any given biological macromolecule, the preceding equations are sufficient to follow the macromolecular path (translations and rotations) with respect to real time. Because of the random term, trajectories must be repeated to obtain an ensemble average or expected behavior.

Anisotropic Diffusion of Biological Macromolecules

Because of the complex structure of biological macromolecules, the diffusion tensor is, in general, anisotropic—that is, it varies along different spatial directions. Whereas isotropic diffusion that is not dependent on direction leads to linear macromolecular trajectories in constant external force fields (and in the absence of the random force), anisotropic diffusion leads to deterministic swirling, twirling, and all sorts of interesting macromolecular trajectories.

As an example, consider a simple, charged peptide (an alpha helix) that has a more linear shape. The deterministic trajectory in a constant external electric field is nonlinear, as shown

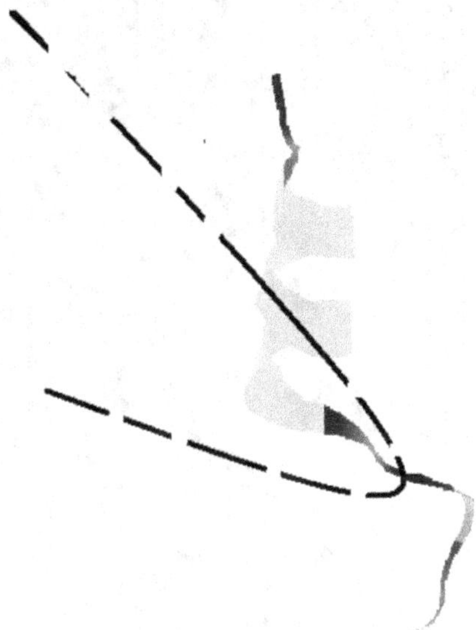

FIGURE 3-8 Illustration of the effects of an anisotropic diffusion or mobility tensor for a macromolecule. The deterministic trajectory shown is the result of motion of a charged, anisotropic alpha helix moving in an electric field, where Brownian motion has been neglected for the purposes of illustration.

in Figure 3-8. In general, biological macromolecules can have preferred pathways and orientations that may play critical roles in their behavior. From a reengineering perspective, it also may be possible to design biological macromolecules in a way that takes advantage of this anisotropic behavior.

Restructuring of Water and Local Water Viscosities

What is equally interesting about diffusion tensor behavior from computer simulations is the unexpected values for charged biological macromolecules. In general, *uncharged* macromolecules with near-isotropic, spherical structure have been shown to compare very favorably with analytical models, which is referred to as *Stokes-Einstein behavior*. A radial distribution function for the solvent, in this case water, characterizes the local

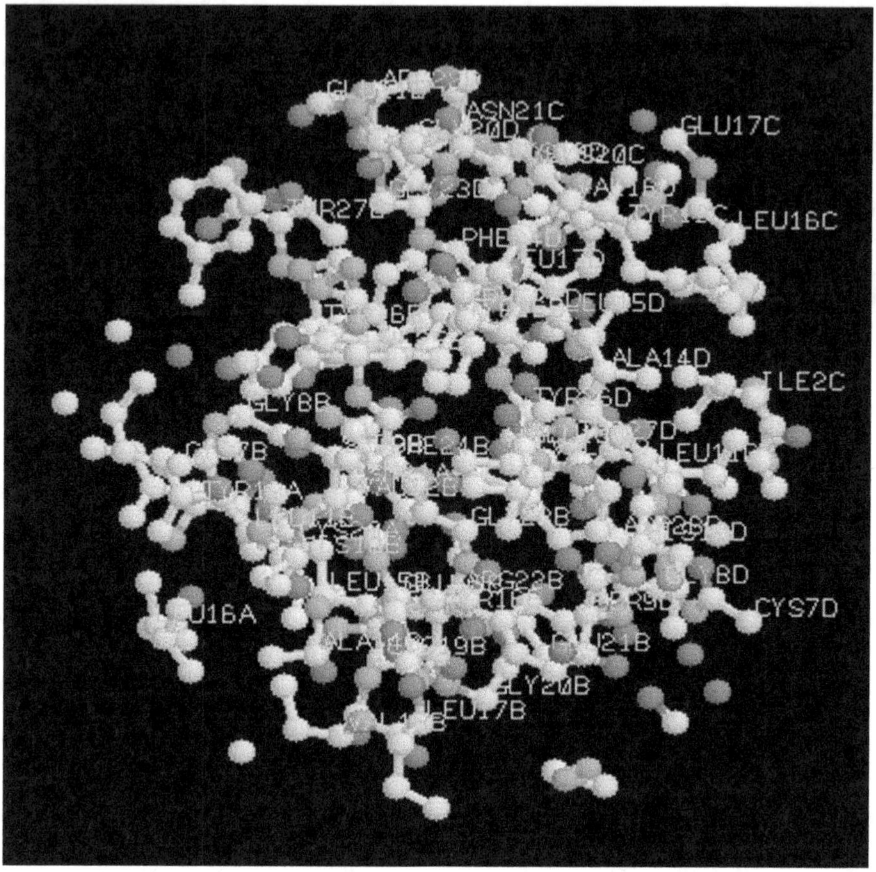

FIGURE 3-9 An artificial spherical protein created to test the Stokes-Einstein expression in biological systems. The accompanying water radial distribution function is shown in the Figures 3-10 and 3-11.

water structure or local density in the vicinity of an isotropic biological macromolecule, as shown here for a "spherical" protein (Figure 3-9). This macromolecule was artificially created by neglecting all atoms from the total protein structure data file (insulin in this case) beyond a prescribed radius, and it allows us to unambiguously study the water microstructure via the radial distribution function, as shown in Figure 3-10.

The peaks shown in Figure 3-10 are known as *coordination shells,* and they represent the high probability of finding water molecules in these layers. Typically, the spacing between peaks

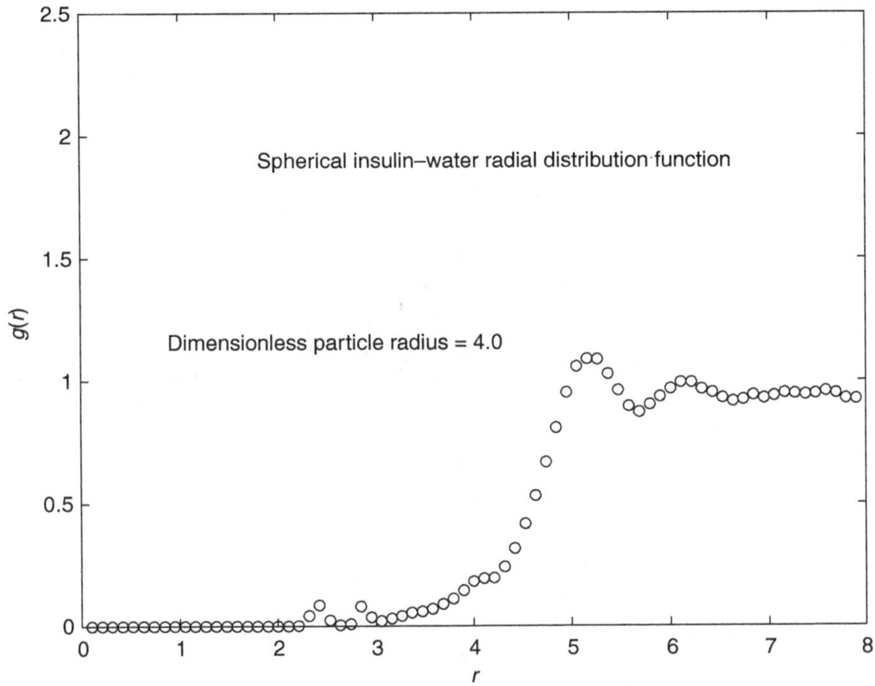

FIGURE 3-10 The water radial distribution function associated with a hydrated spherical protein with the protein charges neglected. This profile is a common one that exhibits peaks or coordination shells representing the high probability of finding a water molecule in these regions.

is on the order of the solvent molecular diameter, and the distribution function decays to unity as distance increases, reflecting purely bulk solvent density. The behavior shown in Figure 3-10 is very typical or orthodox, but we note that the partial atomic charges on the protein have been purposefully neglected in these particular calculations.

However, when partial atomic charges are included, the local water structure and protein diffusion values change significantly. Figure 3-11 shows the same system as in Figure 3-10, but with charges turned on. The coordination peaks completely disappear and are simply replaced by a water-dense, near-field region followed by a water-dilute, far-field region. It is interesting that this behavior was first hypothesized by Hoppert and Mayer (1999) prior to any computer simulations (Figure 3-12).

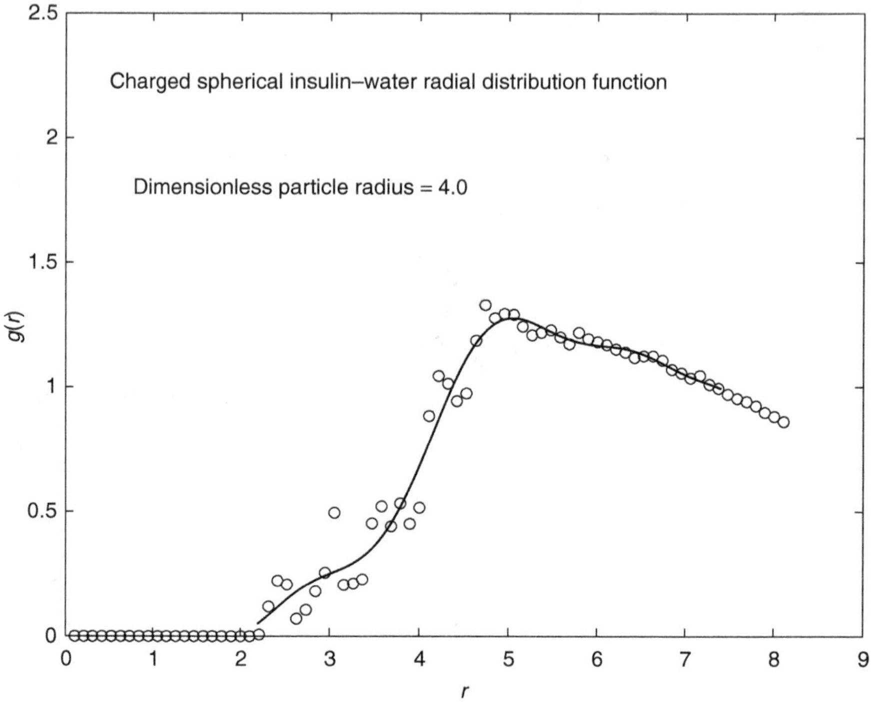

FIGURE 3-11 The water radial distribution function associated with a hydrated spherical protein with the protein charges included. Now, the coordination peaks have disappeared and are replaced by two relatively large layers—that is, a relatively dense water region proximal to the charged surface followed spatially by a relatively dilute region.

In short, the polarity of water results in its pull toward a charged surface; the far-field decreased density is simply the effect of water mass conservation. In addition, the increased density leads to increases in near field local water viscosities over bulk values. This, in turn, translates into decreased diffusion of charged biological macromolecules over what may be expected from size considerations. In fact, experimental measurements of protein diffusion constants or mobilities have shown precisely these types of deviations from Stokes-Einstein behavior.[18]

In general, cells are packed with biological macromolecules, all of which carry some type of charge. The observed gel-like

[18] Lamanna et al. (1997).

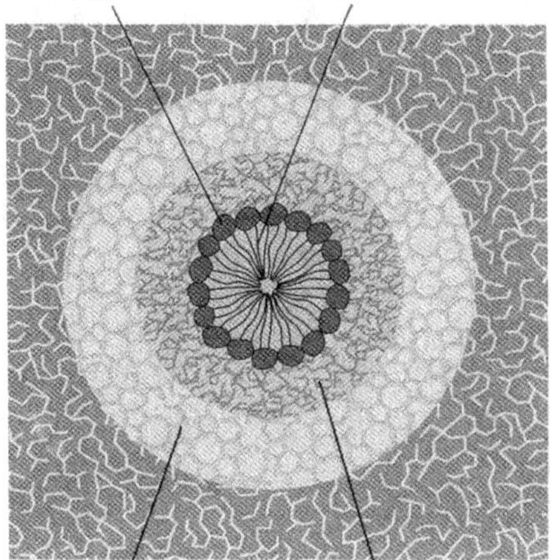

FIGURE 3-12 Illustration of water rearrangement in charged colloidal systems. Adapted from Hoppert and Mayer (1999).

behavior of cell cytoplasm is no doubt the result of water's interaction with these charged species.[19] For example, if any part of the cell plasma membrane is carefully removed or punctured, cytoplasmic fluid does not leak out! Water is tightly bound within cells, and this no doubt has a significant effect on intracellular transport. Much research remains to be done in examining the overall effects of water interactions with charged biological macromolecules, and more is said on this in Chapter 4 when the interactions between macromolecules are presented.

References

Allen, M.P., and Tildesley, D., *Computer Simulation of Liquids,* Oxford University Press, New York (1989).

Berman, H.M., Westbrook, J., Feng, Z., Gililand, G., Bhat, T.N., Weissig, H., Shindyalov, I.N., and Bourne, P.E., *Nucl. Acids Res.,* **28,** 235–242 (2000).

Bratley, P., Fox, B.L., and Schrage, L.E., *A Guide to Simulation,* 2d ed., Springer-Verlag, New York (1987).

Chandrasekhar, S., *Rev. Mod. Phys.*, **15,** 1 (1943).

[19] See Pollack (2001).

Condiff, D.W., and Dahler, J.S., *J. Chem. Phys.*, **44**, 3988–4005 (1966).

Cornell, W.D., Cieplak, P., Bayly, C.I., Gould, I.R., Merz, K.M., Ferguson, D.M., Spellmeyer, D.C., Fox, T., Caldwell, J.W., and Kollman, P.A., *J. Am. Chem. Soc.*, **117**, 5179–5197 (1995).

Cukier, R.I., and Deutch, J.M., *Phys. Rev.*, **177**, 240 (1969).

Goldstein, H., *Classical Mechanics*, Addison-Wesley, Reading, Ma, 1980.

Heermann, D.W., *Computer Simulation Methods in Theoretical Physics*, Springer-Verlag, New York (1990).

Hirschfelder, J.O., Bird, R.B., and Spotz, E.L., *Chem. Revs.*, **44**, 205–231 (1949).

Hoppert, M., and Mayer, F., *Am. Sci.*, **87**, 518–525 (1999).

Lamanna, R., Delmelle, M., and Cannistraro, S., *Phys. Rev. E*, **49**, 5878–5880 (1994).

Kervorkian, J., and Cole, J.D., *Perturbation Methods in Applied Mathematics*, Springer-Verlag, New York (1981).

Lin, C.C., and Segel, L.A., *Mathematics Applied to Deterministic Problems in the Natural Sciences*, MacMillian, New York (1974).

Nayfeh, A., *Perturbation Methods*, John Wiley & Sons, New York (1973).

Papoulis, A., *Random Variables and Stochastic Processes*, McGraw-Hill, New York (1965).

Peters, M.H., *J. Chem. Phys.*, **110**, 528–538 (1999). **3.**, *J. Chem. Phys.*, **112**, 5488–5498 (2000).

Peters, M.H., and Ying, R., *J. Chem. Phys.*, **98**, 6492–6503 (1993).

Pollack, G.H., *Cells, Gels, and the Engines of Life*, Ebner and Sons, New York (2001).

Toukau, K., and Rahman, A., *Phys. Rev. B*, **31**, 2643–2648 (1985).

Tironi, I.G., Sperb, R., Smith, P.E., and van Gunsteren, W.W., *J. Chem. Phys.*, **102**, 5451–5459 (1995).

Van Dyke, M., *Perturbation Methods in Fluid Mechanics*, The Parabolic Press, Stanford, CA (1975).

Zhang, Y., Peters, M.H., and Li, Y., *Proteins*, **52**, 339–348 (2003).

Chapter Four

Dynamics of Interacting Biological Macromolecules

Chapter 3 discussed the movement or dynamics of biological macromolecules in host fluids without regard to their interactions with each other. From a macromolecular functionality point of view, interactions are the rule, and efforts to describe them are critical to analyzing and predicting the cellular processes discussed in Chapter 2, such as cellular growth, metabolism, and so on.

Types and Examples of Biological Macromolecular Interactions

In general, biological macromolecular interactions play a plethora of pivotal roles in each of the following five cellular systems introduced in Chapter 2:

- cell signal processes, signaling cascades, and membrane and transport pathways
- enzymatic systems, or the -ase macromolecules, for biochemical and energy production
- cellular reproduction or division systems
- cellular physical construction, anchoring, and motility
- information systems, checkpointing, and reliability (phenotyping)

All of these systems and associated processes, of course, interplay with one another via biological macromolecular interactions. For example, biochemical production pathways via enzymes are activated by cell signal molecules that produce new molecules for a variety of purposes, such as cell anchoring, other enzymatic activity or cell signaling, and so on. Ultimately, it is this time and space orchestrated interplay that defines overall cellular behavior, and developing quantitative descriptions of each interaction is the first step in building large models of cellular phenomena. Let's look at one important example to illustrate this before proceeding to describe quantitative approaches.

Dynamic Interaction Pathways: Protein Translation

The process of protein production by cellular machinery is depicted in Figure 4-1. Briefly, after proper signaling and preparation, a coding section of DNA is transcribed into messenger RNA (mRNA) in the cell nucleus (RNA synthesis, or transcription stage). The mRNA is transported across the cell nuclear membrane, where translation or protein synthesis takes place in ribosomal units of the cell. In the process of protein translation, two ribosomal units (R1 and R2), consisting of

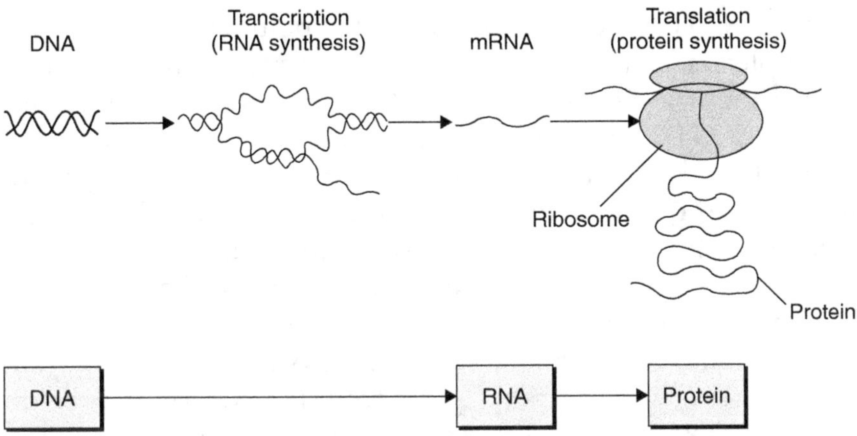

FIGURE 4-1 The basic paradigm of cellular protein production. A DNA coding segment is transcribed into messenger RNA (mRNA). The mRNA complexes with a ribosome unit where protein production takes place (translation).

both polynucleotides and proteins, bind to an initiation site on mRNA along with an initiation factor (IF) complex, as shown in Figure 4-2. The initiation site on mRNA contains the nucleotide sequence AUG, which also is the code for methionine. The IF complex includes a derivatized, complementary transfer

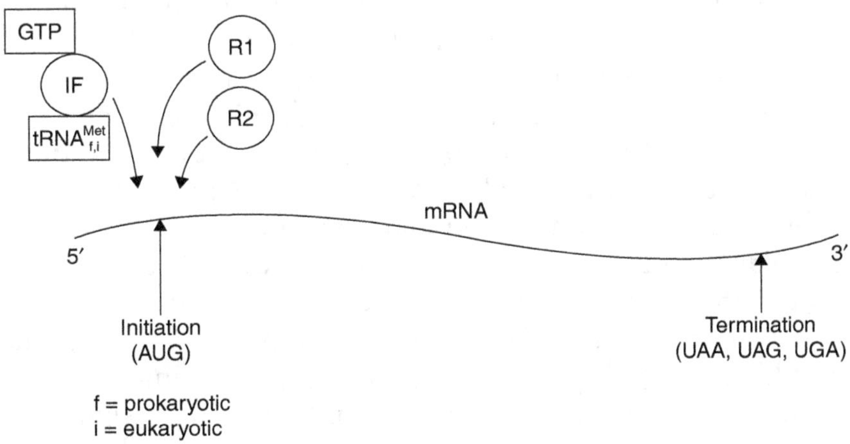

FIGURE 4-2 Details of the complex formation. The ribosome units R1 and R2 combine with an initiating factor (IF) complex. The IF complex contains a type of transfer RNA for methionine that is complementary to the AUG initiation site on mRNA.

RNA (tRNA) for methionine, so that the binding process for the IF complex is of a base-pairing variety (P site binding). Protein translation proceeds with GTP hydrolysis in the IF-bound complex. Since GTP hydrolysis is irreversible, protein translation is unidirectional once it begins. Next, tRNA containing the proper anticodon and associated amino acid binds to mRNA at the ribosome complex site (Figure 4-3, A site binding). If the base pairs match properly, the ribosomal unit alters its configuration, catalyzing the formation of a peptide bond between Met and the amino acid from tRNA at the A site (transpeptidation step). Protein synthesis takes place from the N-terminal region to the C-terminal region of the protein. A translocation step moves the ribosomal unit along the mRNA thread, with the elongated peptide moving to the P site and to the now open A site ready for the next specific tRNA. Peptide elongation proceeds until the termination coding region of mRNA is reached (UAA, UAG, or UGA).

Some important dynamic aspects of this process include the specificity of the binding process, multiple time-orchestrated binding events, the time direction or irreversibility of elongation, and the conformational changes in the ribosomal unit itself. In the process of translation, peptides are produced at an astonishing rate of 20 amino acids per second, and a 200–amino acid protein can be produced in 10 seconds or less. Also of interest is that protein folding, or secondary and tertiary structure formation, begins before the translation termination step.

Clearly, to describe this overall process requires a time-orchestrated, unidirectional, and dynamic analysis of each of the binding and reconformational steps involved. Biomacromolecular interactions, therefore, must be described in a real-time dynamic mode if we are to begin to quantify the overall process of protein translation. This example is typical of the many kinds of cellular process pathways, including those involved in metabolism, DNA replication, intercellular signaling, chemotaxis, and so on. Cells are complex machines that work through the use of chemical and biochemical molecules creating dynamic networks and pathways to accomplish their many orchestrated tasks.

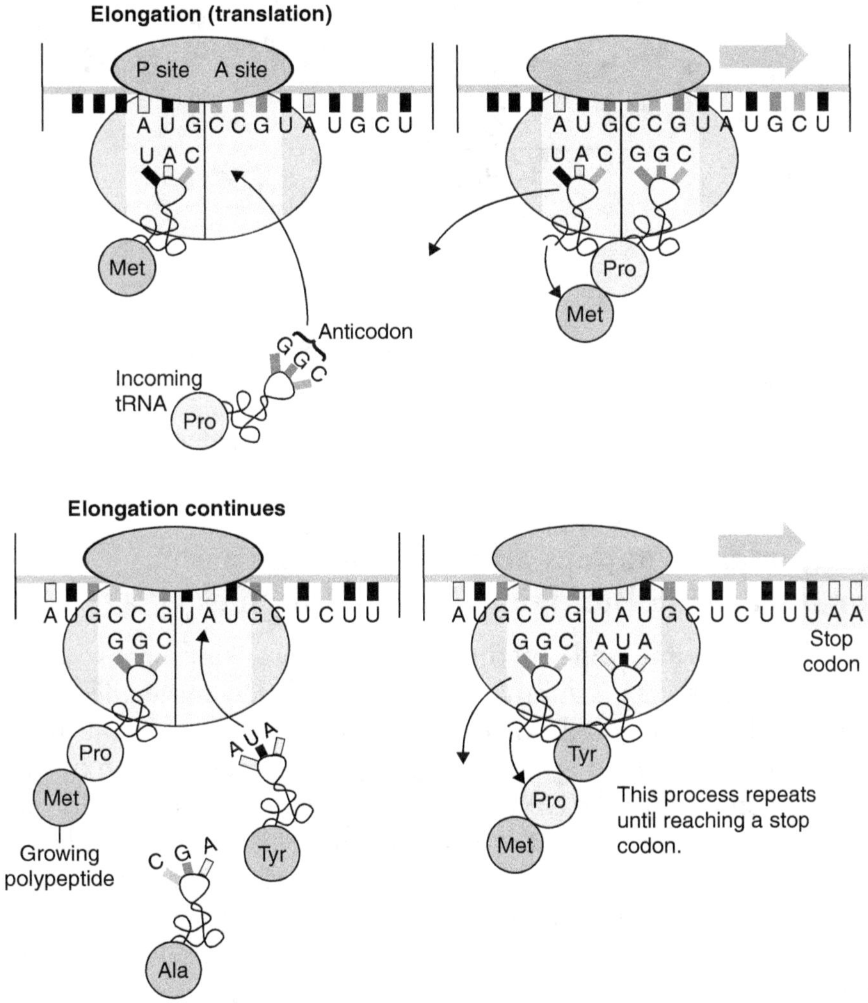

FIGURE 4-3 Details of the protein synthesis process. Complementary tRNA binds to the A site of the complex. The complex catalyzes the formation of a peptide bond between the two adjacent amino acids. Thus, the ribosome complex has enzymatic functions. The addition of amino acids occurs in this chainlike fashion until a stop codon is reached. However, the protein begins assuming its secondary and tertiary structure even before the stop codon is reached.

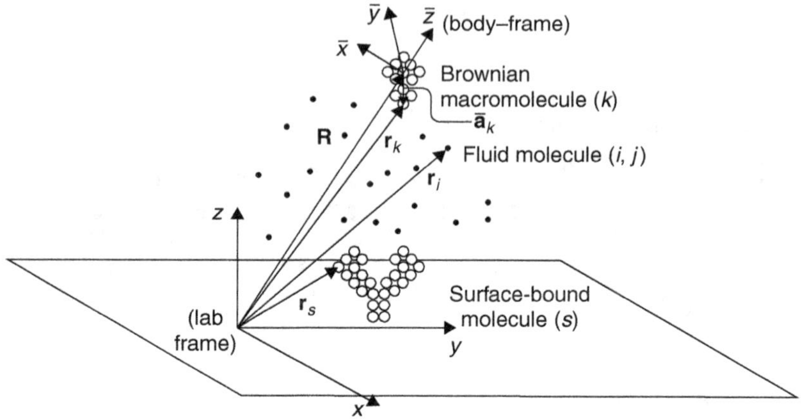

FIGURE 4-4 Variable description of the ligand-receptor binding system. The receptor is considered fixed relative to the mobile ligand, the latter of which can both translate and rotate. The description of rotation requires an identification of a ligand-fixed coordinate system. For a rigid ligand, all ligand atomic positions remain constant relative to the ligand-fixed system.

Molecular Dynamics and Brownian Dynamics for Interacting Biological Macromolecules

The multiple time scale method presented in Chapter 3 to efficiently describe the dynamics of isolated macromolecules in a host fluid is gratifyingly of the same form for interacting biological macromolecules—that is, a displacement equation of the form given can be derived for each biological macromolecule of any particular system.[1] Several variables in the displacement equation, however, have new prescriptions or definitions associated with macromolecular interactions. For future reference, the ligand-receptor system is symbolized and illustrated in Figure 4-4.

The first modification is that the external particle force term now includes the direct particle-particle interaction forces. These direct forces include coulombic interaction forces between charged or partially charged atoms of each macromolecule and Lennard-Jones type force interactions (van der Waals and Born forces), as discussed in Chapter 3. Thus, each

[1] Peters, 1999a; ibid. (2000).

particular atom type of the ligand and receptor must be assigned a charge or partial charge and LJ potential constants to account for these direct forces. The sum rules introduced in Chapter 3 may be used to approximately describe the LJ force between any ligand atom and receptor atom. Note that partial charges include those on hydrogen atoms whose presence may be determined by pH, as discussed in Chapter 1 on proteins. Higher-ordered electrostatic interactions also can be included, such as *induced* atomic dipole forces, but these typically are much harder to characterize.

The second modification has to do with the diffusion tensor. For example, it is well known that to describe the interaction between two or more particles in a host fluid, the diffusion tensor of any one particle is modified by the presence of a second particle or a wall. This effect, called *hydrodynamic interactions,* is normally a diffusional retardation effect—that is, the mobility of any macromolecule is reduced by the presence of a nearby macromolecule or wall.[2] This is a subject that is rich in the field of theoretical fluid mechanics, and no attempt is made here to review all of the different aspects of this area.[3] Ultimately, it is the particular modification of the local fluid pressure and viscous stress fields by the nearby macromolecule, or wall, that leads to increases in local fluid frictional forces acting on the macromolecule and subsequent diffusional decreases via the so-called generalized Einstein relationship:

$$\mathbf{D} = kT\zeta^{-1}$$

where $\mathbf{D}$ here is shorthand notation for the complete, or grand, set of translational-rotational diffusion tensors introduced in Chapter 3, and ζ is the associated macromolecular grand friction tensor.

As discussed in Chapter 3, the macromolecular mobility or diffusion tensor is determined via local equilibrium molecular dynamics and time-force autocorrelation analysis prior to any Brownian displacement calculation. For an isolated biological

---

[2] Ermak and McCammon (1978).

[3] See Brenner (1967), and Happle and Brenner (1983).

macromolecule, this diffusion tensor is invariant in the body-frame of the particle and, therefore, is an intrinsic property of it and need be determined only once. For hydrodynamically interacting biological macromolecules, the diffusion tensor no longer is invariant in the body-frame of the particle and, therefore, must be determined at the beginning of each Brownian displacement via molecular dynamics (MD). Calculations have shown, however, that hydrodynamic interactions in biological macromolecular systems may be relatively less important.[4] First of all, the sizes of biological macromolecules alone, which are on the order of nanometers, can reduce these effects through both weaker alterations in pressure and viscous stress fields for the solvent treated as a continuum fluid—that is, fluids treated from continuum fluid mechanical descriptions, and deviations from fluid continuum behavior itself.

However, even if one were to neglect hydrodynamic interaction effects on the basis of size and other factors, there is still a potential indirect effect of macromolecular interactions hinted at in Chapter 3—that is, the overall gel-like behavior of the cell cytoplasm as a result of charged macromolecular interactions. What may be more important in cellular systems is correctly accounting for the actual solvent viscosity inside cells as far as biological macromolecular mobility is concerned. In the first approximation, one could consider the mobility of any given biological macromolecule as still an intrinsic property of the particle, but, for example, with a solvent viscosity indicative of the actual solvent viscosity inside of a living cell. This would mean appropriately adjusting any diffusion tensor values that were determined via molecular dynamics at some convenient reference solvent viscosity value. By the Stokes-Einstein relation, the diffusion tensor is inversely proportional to the fluid viscosity, which would allow for simple, first-order corrections. We saw in Chapter 3 that charged macromolecules tend to pull polar water molecules toward them, creating proximal, dense water regions and correspondingly distal, dilute water regions, the latter by the need to satisfy water mass conservation. Indeed,

[4] Peters (1999b), and Zhang et al. (2003).

the idea of less viscous internal cell highways, or chreodes, for rapid molecule transport has been suggested by other hydrophobic mechanisms,[5] and there is a clear need for further study of such effects in cells. A spatial viscosity map may provide a pipe-network diagram for the movement of chemicals and macromolecules inside cells. Finally, we note that calculations of the grand diffusion tensor explicitly depend on the equilibrium average force of the fluid acting on the ligand particle as presented in Chapter 3. This force has many complex aspects in interacting biological systems, as discussed more fully here.

A New Force from the Solvent

In addition to potential alterations in the diffusion tensor due to macromolecular interactions, there is also a modification of the net force and torque acting on the macromolecule, the terms $\mathbf{F}'$ and $\mathbf{T}'$, respectively, due to imbalances in the fluid. As is well known, other than the buoyancy force of the fluid acting on a particle in a gravitational field and the so-called phoretic effects such as thermophoresis, the local equilibrium or time-average force of the fluid acting on a single particle is zero. Note that the Brownian force, or random term $\mathbf{C}$, in the Smoluchowski displacement equation is due to particle force *fluctuations* caused by the fluid molecules' incessant bombardment on the macromolecule. This force always is present. For example, you are experiencing it as you read this, but it is simply insignificant for our dynamics because of our extremely large mass compared to that of a fluid molecule.

It is interesting, however, that the local equilibrium or time-average force is not zero for *interacting* macromolecules. In short, for a macromolecule (ligand) near a wall or other macromolecule (receptor), the fluid molecular structure is readjusted in the total potential field it experiences, often creating an imbalance of force on the (ligand) macromolecule. A simple, well-known example is the so-called Asakaru-Oosawa potential in

[5] Waddington (1957), as cited by Kier et al. (2002).

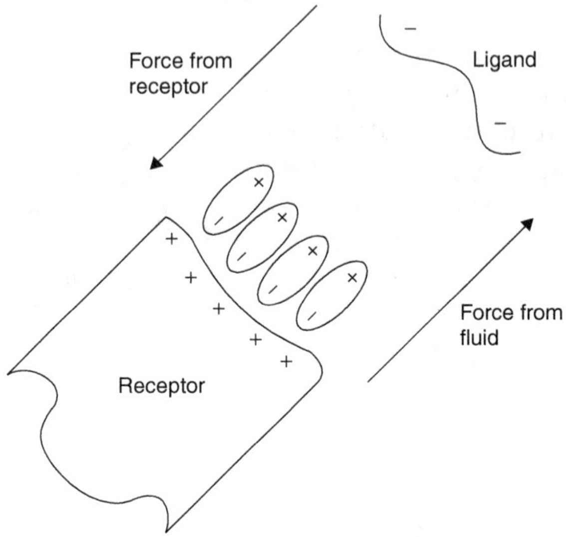

FIGURE 4-5 Simple illustration of the dielectric effect. In looking at the electrostatic force acting on the ligand by the receptor, the presence of a polar solvent will diminish the receptor charge by its alignment along the charged surface of the receptor. In the case of water the effect is highly significant (i.e., it diminishes the force by a factor of 80).

colloidal interactions.[6] As two colloidal particles approach one another at close separation distances in a fluid immersed with polymer molecules, the polymer molecules are unable to gather in the gap space between the two approaching particles. This results in an imbalance of polymer molecules around any colloidal particle, and their bombardment or interactions on the outer particle surfaces pushes the two particles together, creating a net attractive force and particle aggregation.

One of the most important examples, however, of this equilibrium average fluid force acting on a particle is in the analysis of the fluid dielectric constant for interacting charged macromolecules in an electrically nonneutral solvent such as water, as sketched in Figure 4-5. The presence of charge on any macromolecular surface causes polar water molecules to realign according to Coulomb's law. When another charged macromolecule approaches, any of its charged atoms will experience

[6] Asakaru and Oosawa (1958).

a force from the polar fluid molecules that have reoriented themselves. The net effect is a reduction in the coulombic, electrostatic force between the charged atom and charged macromolecule. This reduction is typically quantified by the use of a solvent dielectric constant in Coulomb's law, and this type of equilibrium average fluid force is, therefore, often included *implicitly* in the direct particle-particle interaction force term. However, the overall electrostatic problem in biological systems is more complex due to the ubiquitous presence of ions. Ions immersed in the solvent will also spatially readjust due to the presence of charge on macromolecules in the system. A well-known example is the formation of a double layer of charged ions around colloidal particles and the so-called DLVO (Derjaguin-Landav-Verwey-Overbeek) theory.[7] More generally, for charged, interacting biological macromolecules in ionic solvents, the Poisson-Boltzmann (PB) equation must be solved at each particular configuration in order to correctly account for fluid electrostatic effects.[8]

Alternatively, the molecular dynamics step at the beginning of each BD step allows the exact determination of all fluid electrostatic effects including an accurate accounting of fluid dielectrics and counter ions. As previously noted, there is often a breakdown in the continuum description of the fluid at small interparticle separation distances, and such a breakdown is not considered in the solution to the PB equation. Finally, we note that other effects that are exactly included in the MD step prior to the BD step include so-called desolvation effects, where fluid molecules become displaced by an approaching macromolecule,[9] and the so-called hydrophobic force due to the microstructural reorganization of water molecules around interacting hydrophobic segments of macromolecules.[10]

In summary, then, the force and torque terms appearing in the ligand Brownian displacement equation in Chapter 3 consist of three parts—that is, an external field force or torque if present,

---

[7] See, e.g., Heimenz and Rajagopalan (1997).

[8] See, e.g., the review article by Elcock et al. (2001).

[9] Camacho et al. (2000).

[10] Horton and Lewis (1992).

the direct ligand-receptor interaction force and torque, and the local equilibrium average force and torque of the fluid acting on the ligand,

$$\mathbf{F}' = \mathbf{F}_{\text{ext}} + \sum_k \sum_s \mathbf{F}_{\text{lr}} + \langle \mathbf{F}_f \rangle_{\text{eq}}$$

$$\mathbf{T}' \equiv (T'_\theta, T'_\phi, T'_\psi) = \mathbf{T}_{\text{ext}} + \sum_k \sum_s \mathbf{T}_{\text{lr}} + \langle \mathbf{T}_f \rangle_{\text{eq}}$$

where $\mathbf{F}_{\text{lr}}(\mathbf{r}_s, \mathbf{R}, \Omega; \bar{a}_k)$ and $\mathbf{T}_{\text{lr}}(\mathbf{r}_s, \mathbf{R}, \Omega; \bar{a}_k)$ are the force and torque, respectively, acting on the ligand by the receptor, and the summation extends over all ligand and receptor atoms. The functionality dependency explicitly given here is important in establishing the proper coordinate representation of the force and torque, which in this case represents the force through the center of mass of the ligand and the torque about the center of mass of the ligand, respectively, in the laboratory or fixed frame of the system (Figure 4-4).

Models of Binding Kinetics

Having developed dynamic macromolecular binding behavior from the multiple time scale method, the question becomes how to utilize this information in describing the overall transport and binding dynamics as part of cellular behavior, including physical, experimental analysis of binding events. A simulation strategy is needed that not only includes all details of the ligand-receptor nonequilibrium binding process, but also provides a one-to-one comparison with physical experiments and overall cell modeling approaches. Fortunately, modern multiprocessor computers and programming tools allow for such analysis with relative ease as outlined here.[11] First, we present the basic binding kinetic expressions that commonly are utilized in physical ligand-receptor binding experiments.

In general, a ligand molecule undergoes convective-diffusion in arriving near a target/receptor site. Near the target site, the

[11] Gropp et al. (1999).

ligand experiences the target surface forces as previously discussed. This interaction can result in adsorption or association of particles (binding) to specific sites on the target molecule followed by the possibility of desorption or dissociation of ligand from the receptor surface (Figure 4-6). For generality here, we specifically allow for multiple binding sites on a single target molecule. Target molecules are assumed to be fixed in space, as commonly done in binding kinetic experiments, although the methodology can be easily modified for homogeneous binding experiments. Quantitatively, we can write the binding kinetic expressions in terms of a flux (ligand particle uptake moles per second per unit surface area) for each of these surface interaction processes as

1. adsorption:

$$r_a = ak_a c_A|_{s=0}(\underline{c}_A^{\max} - \underline{c}_A)$$

2. desorption:

$$r_d = ak_d \underline{c}_A$$

In these relationships, k_a [(moles of ligand particles/liter)$^{-1}$s^{-1}] and k_d (s^{-1}) are the rate constants for adsorption and

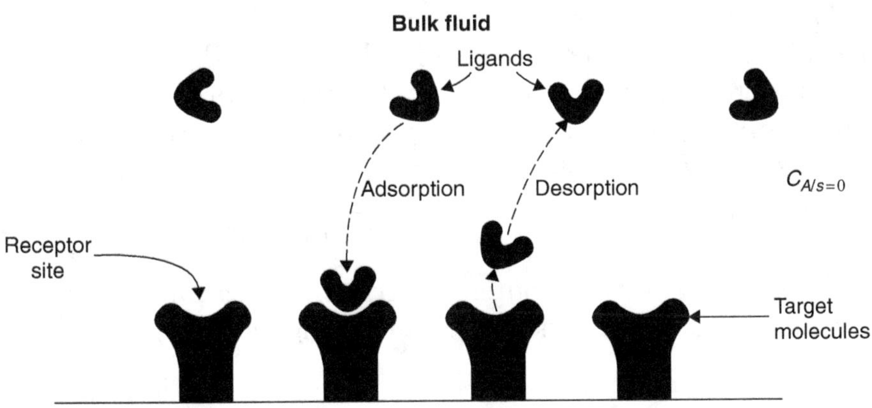

FIGURE 4-6 The various steps involved in the noncovalent binding of a ligand to a receptor or target molecule. These steps include bulk transport to the surface by convection-diffusion, adsorption or binding, and finally desorption from the target.

desorption, respectively, $\underline{c}_A^{max}$ is the (average) maximum number of ligand particles (in moles) adsorbed per target molecule, $\underline{c}_A$ is the *average* actual number of particles (in moles) adsorbed per target molecule, $c_A|_{s=0}$ is the average free ligand particle concentration expressed in M (moles of particles/liter) along the receptor surface ($s = 0$), and a is the number of target molecules per unit surface area. The overall ligand particle surface flux is given, in general, by the boundary condition

$$\mathrm{BC}: N_{A_s}|_{s=0} = -(r_a - r_d)$$

This Langmuir type model is simple in that it ignores any cooperativity effects in the binding process; however, more complex models can be developed along the same lines given here, and in no way does this model present any significant limitations on the computational strategies outlined.[12]

It is important to note that this flux relationship is used as a boundary condition in solutions to the convective diffusion equation for modeling the entire dynamic pathway of ligand transport.[13] More specifically, this modeling approach implicitly represents both an ensemble-average and orientational-average ligand particle behavior.[14] For multiple and distinct binding sites per target molecule, we could write separate flux equations and associated rate constants for each site with separately determined rate constants.

Now, consider a special case of analytical interest—that is, the initial rate regime. At the initial stages of ligand particle exposure to target receptor molecules, the absorbed state concentration is nearly zero—that is, $\underline{c}_A \cong 0$. Under these conditions the flux behavior simplifies to

$$\mathrm{BC}': N_{A_s}|_{s=0} = -ak_a\underline{c}_A^{max} c_A|_{s=0}, \quad \text{initial rate}$$

Both physical and computer experiments measure the total uptake of ligand particles, M_A (moles ligand/time), which is

---

[12] For a more comprehensive description of binding kinetic models, see Lauffenberger and Linderman (1993).

[13] See, e.g., Bird et al. (2002).

[14] Edwards and Brenner (1993).

Table 4-1 Binding Constants*

System	$k_a\,(\mathrm{M^{-1}s^{-1}})$	$k_d\,(\mathrm{s^{-1}})$	K^{-1} (M)
Fibronectin–Fibronectin Receptor	1.17×10^{4}	1.0×10^{-2}	8.6×10^{-7}
Insulin–Insulin Receptor	1.6×10^{5}	3.3×10^{-3}	2.2×10^{-8}
EGF–EGFR	3.0×10^{6}	2.0×10^{-3}	6.7×10^{-10}

*From Lauffenberger and Linderman (1993).

analytically expressed by integration of the previous equation over the total surface to give

$$M_A = N k_a \underline{c}_A^{\max} c_A|_{s=0}, \quad \text{initial rate}$$

where N is the total number of target molecules. We also note that, on the other extreme, by allowing the system to reach a steady state, we have

$$N_{A_s}|_{s=0}(t=\infty) = 0$$

and, thus, from the general flux expression

$$\underline{c}_A(t=\infty) = \frac{K\underline{c}_A^{\max} c_A|_{s=0}}{1 + K c_A|_{s=0}}, \quad \text{steady state}$$

where the equilibrium constant $K(\mathrm{M}^{-1}) = k_a/k_d$. This equation is the Langmuir surface adsorption isotherm for the system, and, as is well known, it can be rearranged to construct a so-called Scatchard plot to determine K and $\underline{c}_A^{\max}$ from steady-state surface concentration data.

According to this last relationship, half of the receptor sites will be occupied when the free ligand concentration is equal to K^{-1}. Table 4-1 shows some typical values in biological ligand-receptor systems, where nanomolar values are indicative of relatively strong binding systems and micromolar values represent relatively weaker binding behavior. As a benchmark, the avidin-biotin system has a K^{-1} of around 10^{-15}M, which is the one of the strongest known noncovalent biological binding systems.[15]

Now, it also is instructive in the overall analysis to nondimensionalize and scale these equations by introducing the following

[15] Green (1975). Also, see Grubmuller et al. (1996).

dimensionless variables:

$$
\begin{aligned}
N_A^* &= N_A/(D_A c_{A_0}/l_0) \\
c_A^* &= c_A/c_{A_0} \\
\underline{c}_A^* &= \underline{c}_A a/c_{A_0} l_0 \\
s^* &= s/l_0
\end{aligned}
$$

where D_A is taken as the scalar translational diffusion coefficient for the ligand particle $[Tr(\mathbf{D}_T)]$ and l_0 is a characteristic macroscopic length scale.[16] In dimensionless terms, the surface boundary condition becomes

$$\mathrm{BC}(3): N_{A_s}^*|_{s=0} = \phi_a c_A^*|_{s=0}(\underline{c}_A^{\max *} - \underline{c}_A^*) - \phi_d \underline{c}_A^*$$

where $\phi_a = k_a l_0^2 c_{A_0}/D_A$ and $\phi_d = k_d l_0^2/D_A$ are known as *Thiele moduli* for adsorption and desorption, respectively. These dimensionless groups characterize the ratio of the rate of adsorption and desorption, respectively, to that of ligand diffusion. In general, small values of the Thiele modulus indicate adsorption or desorption is controlling, whereas large values indicate a diffusion controlled system. As an example, consider a protein ligand with a diffusion coefficient of 10^{-6} (cm^2/s), a reference concentration of 1μM, and a macroscopic length scale of 10 nm. For k_a of 10^6 ($\mathrm{M}^{-1}\mathrm{s}^{-1}$), the Thiele modulus for adsorption is on the order of 10^{-6}, indicating an adsorption controlled system—that is, the rate limiting step is adsorption and not diffusion.

Figure 4-7 shows the basic simulation box for the mimic biophysical kinetic binding experiment. The center of mass of a single target receptor is placed at the center of the box, and a fixed lab frame Cartesian coordinate system is assigned to it. Ligands are initially placed randomly within the available volume of the box according to a given fixed initial concentration. The available volume excludes the box volume occupied by the receptor, and the center of mass of any ligand must reside within the box

[16] See Peters (2005), Chapter 6, for a more detailed discussion of scaling and dimensional analysis.

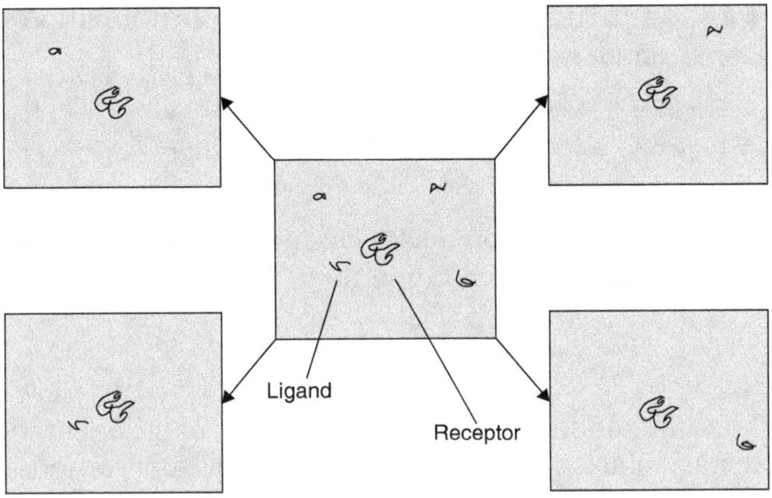

FIGURE 4-7 A simulation box for the computation of ligand-receptor binding. Periodic boundary conditions are used along the box boundaries. Ensemble averages are performed by assigning random initial positions and orientations of the ligands. For noninteracting ligands, the overall simulation may be partitioned as shown, making it highly amenable to grid or parallel processing.

and not outside. In addition to the random assignment of the center of mass for each ligand, the ligands also are assigned random orientations with respect to the lab frame through a random assignment of their respective Euler angles. For the ligand molecules, the known invariant body-frame atomic coordinates and initial body-frame ligand grand diffusion tensor, the latter of which is determined from MD and force autocorrelation analysis, are transformed to the laboratory frame using the initial assigned Euler angles.[17] In particular, if $A_{\bar{j}}$ is any vector component in the body frame of the ligand particle, the component A_i in the lab frame is given in terms of the Euler angles (θ, ϕ, ψ) by

$$A_i = \sum_{\bar{j}} e_{i\bar{j}} A_{\bar{j}}$$

[17] Goldstein (1980).

where $e_{i\bar{j}}$ is a particular component of the transformation matrix given in total by

$$[e_{i\bar{j}}] = \begin{bmatrix} \cos\phi\cos\psi - \sin\phi\cos\theta\sin\psi & -\cos\phi\sin\psi - \sin\phi\cos\theta\cos\psi & \sin\phi\sin\theta \\ \sin\phi\cos\psi + \cos\phi\cos\theta\sin\psi & -\sin\phi\sin\psi + \cos\phi\cos\theta\cos\psi & -\cos\phi\sin\theta \\ \sin\theta\sin\psi & \sin\theta\cos\psi & \cos\theta \end{bmatrix}$$

In the case of the diffusion tensor components, the transformations are applied twice according to

$$D_{ij} = \sum_{\bar{k}}\sum_{\bar{l}} e_{i\bar{k}} e_{i\bar{l}} D_{\bar{k}\bar{l}}$$

Note that the random assignment of ligand initial orientations is necessary since the kinetic binding model expressions are orientational-average quantities as previously noted.

After this initial assignment, each ligand is moved according to the Brownian *coupled* translational-orientational displacement equations given in Chapter 3. Now, as we have previously discussed, hydrodynamic interactions, which lead to *position-dependent* ligand orientational and translational diffusion, often are of negligible importance in small ligand systems. Also, as just noted, simulations often neglect, as a first approximation, any changes in the fluid equilibrium average force acting on the ligand due to water microstructural changes with ligand position changes (i.e., hydrophobic forces and structure-based dielectrics). Under these conditions, it is necessary to perform the computationally costly MD only once prior to any BD steps to obtain the now invariant lab frame grand diffusion tensor. As previously noted, when ions are present in the solvent, a PB step or some other electrostatic model is needed to determine the net electrostatic force, including dielectrics, acting on the ligand at any time.[18] But, generally, a PB step is far less computationally costly than a MD step.

In the more general case, where the particular fluid microstructure is considered throughout the simulation, the ligand-receptor system in its new configuration must be rehydrated

[18] See, e.g., the review by Elcock et al. (2001).

prior to the next BD step and the system allowed to reach local fluid phase equilibrium through a new MD calculation.[19] The new grand friction tensor and new fluid phase equilibrium force are determined in the MD step, and they are subsequently used as input parameters in the next BD step. In general, the MD time step is on the order of femtoseconds with a typical fluid relaxation simulation time on the order of picoseconds, whereas the BD time steps are on the order on nanoseconds with a typical ligand association time on the order of microseconds. Also, whereas the MD steps require $6N$ dimensions (N equal to the total number of system atoms including ligand, receptor, and solvent), the BD step is a mere six dimensions (three translational and three rotational degrees of freedom for the ligand particle as a rigid, atomic structured particle). The entire process MD followed by BD is repeated as the ligand trajectory is traced out in real time. The time steps in the BD calculation are set by the diffusion limit condition, as discussed in Chapter 3 (also, see the next example).

Finally, then, the ligand positional changes are followed in time using standard periodic boundary conditions along the box walls. For simplicity, the size of the simulation box is chosen such that the ligand-receptor interaction force along the box walls is negligibly small, albeit we note that other schemes to reduce the box size have been proposed in the past,[20] which perhaps may be of less importance in view of modern computers and grid processing capabilities. During the simulation the ligand trajectories are monitored to determine adsorption or association with the target receptor, and this is readily done through monitoring of the total ligand-receptor interaction potential as will be shown in the example.

For the *initial* rate model studies, we can ignore the presence of bound ligands on any particular ligand transport and binding trajectory—that is, all ligands see a "bare" target receptor. The computational scheme is, thus, highly amenable to multiprocessor applications where each processor computes the

[19] Peters (2000).
[20] Luty et al. (1992).

trajectory of a single ligand particle interacting with a single target molecule. The trajectories and number of adsorbed ligand particles across all processors are recorded with time to obtain ensemble average ligand concentrations and the net surface flow of particles.

Here we demonstrate details of the algorithm on a specific system showing particular computational values that are necessary to effect a complete mimic experimental simulation, as well as specific details for determining the kinetic binding constants.

Example: gp41/D-peptide Receptor/Ligand System

Pathologically important ligand-receptor systems include viral capsid and envelop protein attachments to cell surface proteins as entry points to the cellular machinery necessary for viral replication and numerical growth patterns.

Here we examine the kinetic binding properties of a particular D-peptide ligand (proposed inhibitor drug) and its interactions with a segment of an HIV capsid protein called gp41. The complete molecular structure of the D-peptide/gp41 bound complex is available[21] (see Figure 4-8) and is utilized to exemplify and illustrate the mimic biophysical computer experimentation previously outlined. We note that the gp41 pocket, or target site in this case, is of a hydrophobic variety due to the particular nonpolar amino acids comprising it. Hydrophobic binding segments are known to be important in viral systems, since they often represent conserved regions during viral mutations.

The D-peptide ligand was first separated from the complex and examined in isolation through MD for its intrinsic grand diffusion tensor. The complete diffusion tensor, details of the initial MD step, and other specific input data (charge, Lennard-Jones force constants, etc.) have been described previously and are not repeated here.[22]

Next the gp41 target was placed in the center of a simulation box with the α helix axis along one of the three Cartesian lab

---

[21] Eckert et al. (1999).

[22] Zhang et al. (2003). Also, see Chapter 3 for a discussion of force constants.

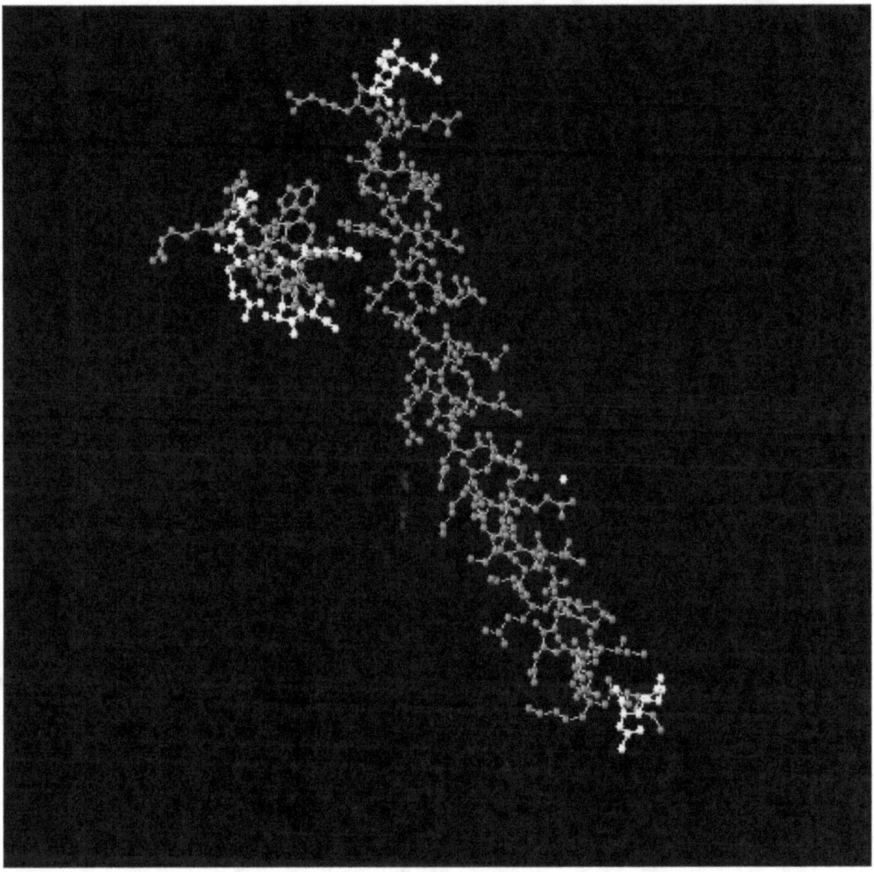

FIGURE 4-8 The gp41/D-peptide complex used in the example simulation study. Molecular structure information for both the receptor and ligand, via X-ray crystallography or NMR, are necessary before any simulation can be done. The system must also be hydrated at the beginning of any molecular dynamics step.

frame directions. A box size of around 4000 nm^3 surrounding the hydrophobic pocket was used based on some initial trial-and-error simulations of the net interaction forces. Ligands were placed randomly, both in position and orientation with their centers of mass inside the control volume. Because of the assumption of noninteracting ligand particles (physically dilute), ligand trajectories can be analyzed independently of one another and are, thus, subject to efficient grid processing or multiprocessing calculations. For example, if there are L total ligands and K total processors, one can simulate L/K ligands per processor.

The analytical, initial rate regime expression previously given can be used for guidance as to the total number of ligands necessary in any simulation. In particular $(k_a c_A|_{s=0})^{-1}$ represents the average time for a single ligand to associate with a single target site. For a diffusional distance of around 10 nm in this case (based on box size), a time step of 10^{-11} s, and for the particular diffusion coefficient estimated for the D-peptide, 1000 total time steps are minimally needed for a single ligand to transverse the simulation box. Thus, for a total trajectory time of 0.1 μs, according to the initial rate expression approximately 100,000 ligand trajectories are minimally required in strong binding systems (k_a around 10^6 $M^{-1}s^{-1}$), and 10 M would be minimally required in a weak binding system (k_a around 10^4 $M^{-1}s^{-1}$). Thus, for a 100-processor computer, anywhere from 1000 to 100,000 ligands per processor are required.

The ligand trajectories in the near bound state of the target segment were isolated through examination of the total potential interaction and interatomic separation distances between the ligand and receptor. As shown in Figure 4-9, the total interaction potential, which is dominated by short-range van der Waals and Born forces, correctly demonstrates a long-term oscillatory behavior indicative of a particle trapped in a potential well. Interatomic van der Waals attractive forces, which pull ligand atoms toward the receptor, are counterbalanced by Born repulsive forces. Such oscillatory behavior is a clear indication of a bound state in hydrophobic binding systems, and analysis of ligand trajectories over each member of the ensemble easily can be made in order to determine the number of bound state ligands at any time. Note that in the near bound state, a variable time step is needed due to the large forces involved. To accomplish this, the time step for any displacement calculation is selected to ensure a macromolecular rotation of only a few degrees, or about 0.01 radians per time step. From the previous Brownian displacement equation, the variable time step is

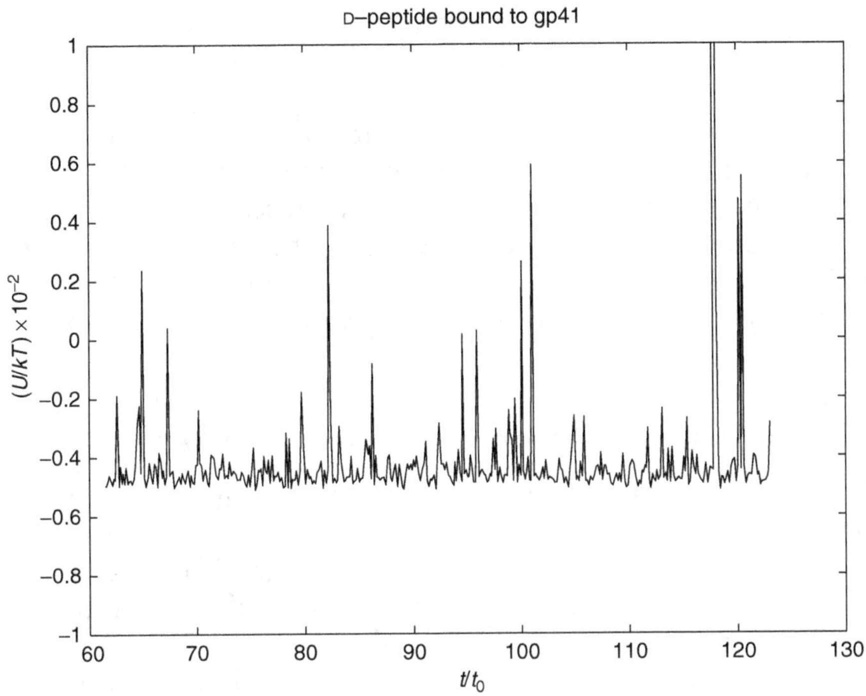

FIGURE 4-9 The potential energy "signature" of binding. As the ligand binds to the surface strong repulsive forces keep it from penetrating, whereas strong attractive forces keep it from leaving. Thus, the ligand vibrates at the target site locked in to a chaotic, Brownian orbit. A large burst of fluid kinetic energy (Brownian force) can kick the ligand out of the target site.

estimated from the values of the trace of the rotational diffusion tensor and torque as

$$t = \frac{0.01}{\epsilon \mathrm{Tr}[\mathbf{D}_R^0] T^0}$$

where $\mathrm{Tr}[\mathbf{D}_R^0]$ is the trace of the rotational diffusion tensor, and T^0 is the modulus of the torque. The time step throughout the simulation is then taken to be the minimum of the initial fixed time step, which is based on no interaction forces, and the variable time step calculated from this expression. We note that in the bound state, it often is difficult to maintain a diffusion limit, because the characteristic force length scales R_0 are very small, as discussed in detail in Chapter 3. A more accurate bound state simulation requires a Fokker-Planck description of the macromolecular dynamics.[23]

Finally, we note that the large number of ligand trajectories required in this particular simulation (approximately 1 M for a single binding event to the particular target segment) are indicative of a intermediate binding regime for this particular ligand. This was also confirmed by the literature report of μM experimental values for the inverse binding equilibrium constant.[24]

The Importance of Conformational Changes

All of the analysis presented up to this point was for structured, rigid ligands and receptors. As noted in Chapter 2, many of the interesting and important features of ligand-receptor interactions are related to changes in macromolecular structure as a result of binding. Intramacromolecular conformational changes help propagate specific signal pathways, and they represent the heart of enzymatic actions. As we saw in the example in this chapter, ribosomal units have multifunction job tasks, all of which depend on conformational flexibility: they function as the physical foundation in the cell signal cascade and act as an enzyme (a peptidyl transferase) in protein translation. Macromolecular conformational changes add to the high degree of specificity and time-space orchestration of cell signal pathways and cellular functions in general.

The simplest way to add flexibility to the MD/BD analysis just given is to identify rigid, Brownian subunits for a given biological macromolecule (ligand or receptor). The intramacromolecular interactions (or interactions between the subunits of a particular macromolecule) that result in conformational changes can be expressed in terms of a generalized joint potential that accounts for all possible degrees of freedom at the joint or point of interaction between adjacent subunits.[25]

[23] Peters (1999a).

[24] Eckert et al. (1999).

[25] Peters and Ying (1993).

Alternatively, specialized joints or degrees of freedom particular to any macromolecule can be used as, for example, the peptide Ramachandran angles introduced in Chapter 1. Peptide bonds in proteins manifest into the rigid, planar-shaped subunits, and the interactions between the planar subunits can be described in terms of the Ramachandran angles. Flexible DNA and RNA also have been modeled in a similar manner.[26]

Theoretically, however, there are a number of precautions necessary in breaking up any macromolecule into rigid interacting, Brownian subunits. Although the mass of any subunit often is much larger than the mass of a single solvent molecule (the basic Brownian assumption), the Smoluchowski diffusion limit assumption may not always hold due to the magnitude of the interaction forces between subunits, analogous to the bound state simulation remarks previously made. Again, the more general Fokker-Planck description may be necessary for accurate simulations.

References

Asakura, S., and Oosawa, F., *J. Polym. Sci.*, **33**, 183–192 (1958).

Barkley, M.D., and Zimm, B.H., *J. Chem. Phys.*, 70, 2991–3007 (1979).

Bird, R.B., Stewart, W.E., and Lightfoot, E.N., *Transport Phenomena*, 2d ed., Wiley, New York (2002).

Brenner, H., *J. Coll. Interface Sci.*, **23**, 407 (1967).

Camacho, C.J., Kimura, S.R., DeLisi, C., and Vajda, S., *Biophys. J.*, **78**, 1094–1105 (2000).

Edwards, D., and Brenner, H., *Macrotransport Processes*, Butterworth-Heinemann, New York (1993).

Eckert, D., Nalashkevich, V., Hong, L., Carr, P., and Kim, P., *Cell*, **99**, 103–115 (1999).

Elcock, A.H., Sept, D., and McCammon, J.A., *J. Phys. Chem. B*, **105**, 1504–1518 (2001).

Ermak D.L., and McCammon J.A., *J. Chem. Phys.*, **69**, 1352–1360 (1978).

Goldstein H., *Classical Mechanics*, Addison-Wesley, Reading, MA (1980).

Green, N.M., *Adv. Protein Chem.*, **29**, 85–133 (1975).

Gropp W., Lusk E., and Skjellum A., *Using MPI*, The MIT Press, Cambridge, MA (1999).

Grubmuller, H., Heymann, B., and Tavan, P., *Science*, **271**, 997–1000 (1996).

Happel, J., and Brenner, H., *Low Reynolds Number Hydrodynamics*, Nijhof, Dordrecht, The Netherlands (1983).

Hiemenz, P.C., and Rajagopalan, R., *Principles of Colloid and Surface Chemistry*, 3d ed., Marcel Dekker, New York (1997).

[26] Barkley and Zimm (1979).

Horton, N., and Lewis, M., *Protein Sci.,* **1,** 169 (1992).

Kier, L.B., Cheng, C.-K., and Testa, B., *J. Theor. Biol.,* **215,** 415–426 (2002).

Lauffenberger, D.A., and Linderman, J.L., *Receptors: Models for Binding, Trafficking, and Signaling.* Oxford University Press, New York (1993).

Luty, B.A., McCammon, J.A., and Zhou, H.-X., *J. Chem. Phys.,* **97,** 5682–5686 (1992).

Peters, M.H., *J. Chem. Phys.,* **110,** 528–538 (1999a).

———., *J. Stat. Phys.,* **94,** 557–573 (1999b).

———., *J. Chem. Phys.,* **112,** 5488–5498 (2000).

Peters, M.H., and Ying, R., *J. Chem. Phys.,* **98,** 6492–6503 (1993).

Peters, M.H., *Molecular Thermodynamics and Transport Phenomena: Complexities of Scales in Space and Time,* McGraw-Hill, New York (2005).

Zhang, Y., Peters, M.H., and Li, V., Proteins: Structure, Function, and Graphics, **52,** 339–348 (2003).

Chapter Five

Toward a Whole Cell Description

In Chapter 4, we developed a real-time molecular-level description of ligand-receptor interactions that included complete molecular details of the structure and dynamic interactions of biomacromolecules and their associated solvent. In considering the thousands of biomolecular interactions that take place in cells over relatively large spatial regions and long time periods, it is not fathomable that a complete brute-force molecular dynamic simulation of the living cell would be possible. Indeed, we demonstrated that such an approach is not even necessary due to the inherent disparate time scales of the underlying physicochemical phenomena. The same disparities are true for spatial scales, as is further discussed in this final chapter.

From the Microscopic to the Macroscopic Description

Chapter 4 also demonstrated that multiscale, molecular-based (microscopic) analyses can be translated into a larger-scale, average behavioral description, or macroscopic analysis. In particular, we were able to cast molecular-based analyses of the real-time dynamics of ligand-receptor interactions into much simpler dynamic rate equations involving macroscopic species concentrations and associated macroscopic rate constants of the binding process. The translation from the microscopic description to the macroscopic was accomplished through the process of so-called ensemble averaging—that is, the conducting of simulations and analyses over all of the possible initial realizations, or configurations, of the system. Ensemble averaging is at the heart of the field of statistical mechanics that has as part of its general mission to relate the microscopic world to the macroscopic.[1]

The goal of the theoretical analysis of cellular subsystems and their integration in the living cell is to develop the correct macroscopic dynamic description and analysis from the underlying molecular or microscopic behavior for each and every type of pathway involved in cellular phenomena. Cellular pathways are both space and time dependent, where the spatial dependency is associated with the physical structure of the cell and its constituents. So, we begin by considering the various types of dynamic pathways and associated cellular spatial organizations, or cellular anatomy.

Cellular Anatomy and Pathway Analysis

The living cell is a system of high spatial organization, or low entropy. A typical cell and its cellular contents are illustrated in Figure 5-1. Internal spatial organization is a critical part of cellular function. The endoplasmic reticulum involved in protein translation, for example, is proximal to the cell nucleus and

[1] See, e.g., McQuarrie (1976).

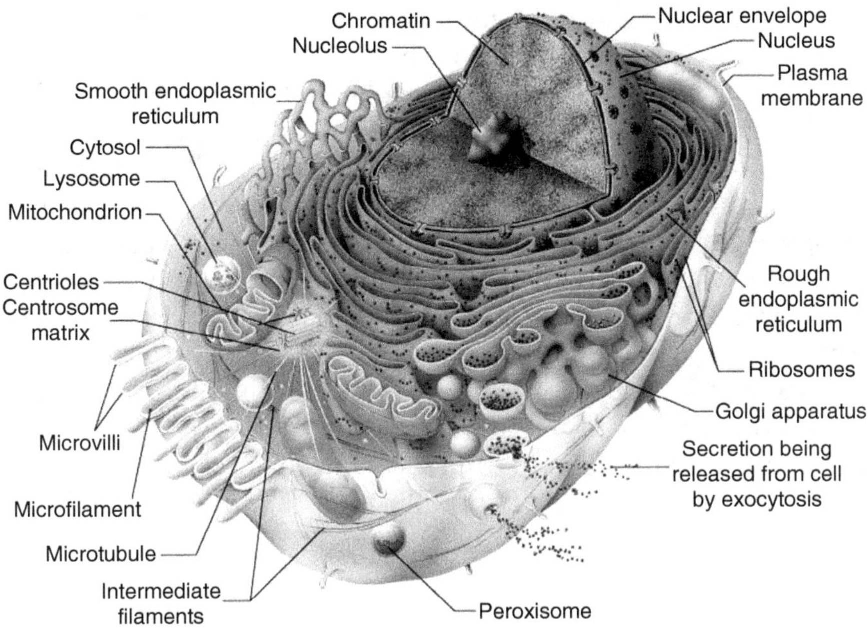

FIGURE 5-1 Generalized description of the living cell and its subcellular contents. Differentiated cells vary widely in terms of their specific elements and functions. For example, erythrocytes contain no cell nucleus (anucleated). The plasma membrane is a more fluidlike structure, allowing for restricted planar diffusion of its components (fluid mosaic model). The cytoplasm is also more gel-like than fluidlike owing to the highly packed and highly electrically charged nature of its contents. Each cell can be envisioned as a highly complex biochemical factory. From Elaine N. Marieb, *Human Anatomy and Physiology,* Pearson Publishing, by permission.

the transcription machinery. The anisotropic nature of cellular contents provides important information about the cell's position and orientation in its associated tissue, imparting a sort of proprioceptive ability to the living cell. Indeed, the anisotropic behavior of embryonic stem cells and their progenitor cells, for example, are well known to precisely direct embryological development.[2] The semistatic positioning of cellular contents is possible, in part, because of the gel-like behavior of cell cytoplasm, as noted in Chapter 3,[3] and the actin structural elements discussed in Chapter 2.

[2] Sadler (2004).
[3] Pollack (2001).

In order to simulate the whole cell, we must integrate the subsystem behavior introduced in Chapter 2 (cellular physiology, or cellular function) with the particular cellular anatomy, or internal physicochemical structure, of the living cell. To do this, we introduce a set of pathways associated with cellular function and link those to their associated spatial segments of the cell.

Figure 5-2 depicts a highly simplified pathway network diagram for the whole cell. The pathways are associated with the subcellular systems, such as the pathways for *transport systems, signal transduction systems, biochemical production systems,* and so on. The motility and mechanical work subsystem is contained in the term W, the rate of work done by the cell on its surroundings. However, we purposefully draw the arrow two ways in Figure 5-2 to illustrate that the surroundings

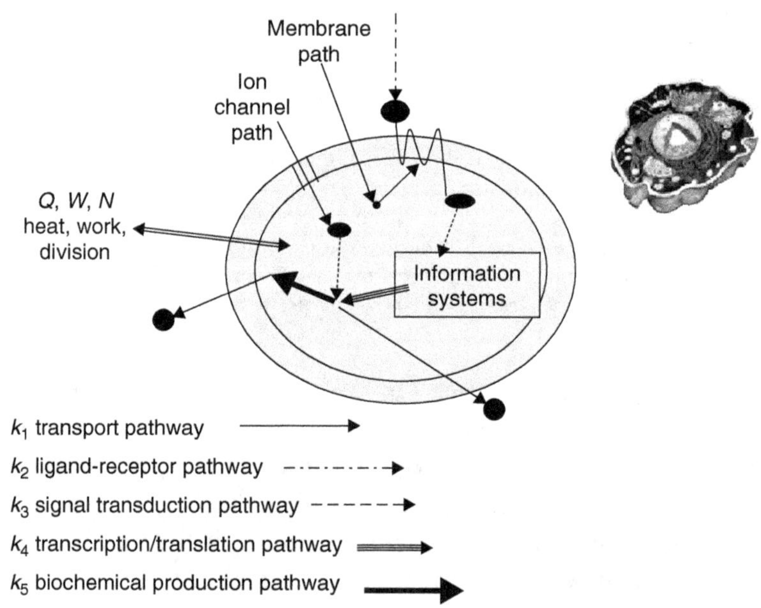

FIGURE 5-2 Simplified pathway network diagram for the living cell. Extracellular molecules can interact with the cell in a variety of ways, including passive diffusion of small hydrophobic substances across the plasma membrane, transport through membrane channels, and interactions with integral membrane proteins. The various pathways are associated with each of the subcellular systems introduced in Chapter 2.

could do work on the cell, such as extracellular shear forces and stresses. Cells often contain stress receptors in their plasma membranes, as, for example, in the cochlea of the inner ear. Other examples include smooth muscle cells, where excess external stresses can lead to smooth muscle cell proliferation. Another important feature of the cell interaction with its environment is the transfer of energy across the cell boundary, also explicitly indicated in Figure 5-2. The rate constants associated with each pathway, notably the biochemical production pathway, can be highly temperature dependent, and energy exchange between the cell and its environment helps regulate particular pathways. Change in cell number via the reproduction subsystem machinery, or in the opposing direction of cell death, such as through apoptosis, also is included in Figure 5-2. Nutrients and waste products enter and leave the cell, respectively, via the membrane transport pathways, and manufactured proteins can leave via lipid packaging and exocytosis. Now, all pathways involved in a particular function of the cell must be mathematically described in both space and time in order for us to proceed with their integration into overall cell behavior.

Dynamic Rate Equations and Pathway Integration

In this section, we consider a highly simplified pathway analysis and integration in order to illustrate the essential features of a comprehensive approach to quantify cellular behavior. As previously emphasized, the pathways associated with a particular subcellular system, such as reproductive/division system, biochemical production system, and so on, must be identified and described in an explicit temporal and spatial dependent manner in order to eventually integrate each subsytem into an overall description of cellular behavior. In general, there may be thousands of pathways involved in an accurate description of each subcellular system. Many, if not most, internal cellular pathways have yet to be described in any fundamental, molecular-based mathematical detail, with the exception of the field of cellular metabolism or the biochemical production system.[4]

Let's consider, as an example, a simple process of the cellular biochemical production of an enzymatic product compound involving two subcellular systems, (1) external/internal communication and transport system and (2) enzymatic and biochemical production system. In general, each subsystem may involve a multitude of interconnecting pathways, but here we consider just a few in order to illustrate the essential mathematical features of the analysis.

In this simple example, an external cell signal molecule binds with a membrane target that leads to the release of a second messenger molecule. The second messenger molecule must undergo transport to a particular cellular region where it initiates an enzymatic reaction to produce a particular biochemical product molecule. The first pathway considered, therefore, is an external cell signal communication pathway that we have treated in some detail in Chapters 3 and 4. As shown in Figure 5-3, a signal molecule *A* from some source appears in the extracellular region and binds to a surface receptor molecule. We are not concerned with the description of the extracellular processes that resulted in the particular appearance of the signal molecule, although note that the basic principles and techniques introduced here also can be applied to quantify and analyze extracellular pathways.

The rate of change of the ligand species *A* is given by a simple mass balance at the cell surface as

$$
\begin{aligned}
\frac{d\underline{c}_A}{dt} &= -SN_{A_s}|_{s=0} \\
&= S[ak_a c_A|_{s=0}(\underline{c}_A^{\max} - \underline{c}_A) - ak_d\underline{c}_A]
\end{aligned}
$$

where all variables were defined in Chapter 4. This relationship describes the time rate of change of the adsorbed species concentration in terms of the adsorption and desorption rate constants that are determined from the associated molecular analysis and ensemble averaging methods.

[4] See, e.g., Heinrich and Schuster (1996).

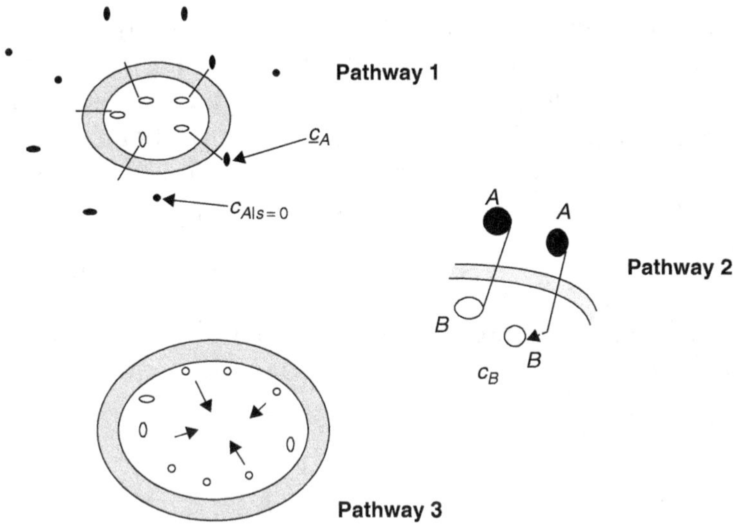

FIGURE 5-3 Simple, four-pathway sequence for quantitative analyses. The first pathway involves a cell signal molecule (ligand) binding to an integral membrane protein (receptor). This induces the release of a secondary signal molecule (pathway 2), followed by its cellular diffusion (pathway 3). Finally, the signal molecule induces a localized enzymatic reaction (pathway 4; not depicted in the figure).

Now, as noted previously, the magnitude of the Thiele modulus in ligand-receptor processes is often such that one can safely neglect any mass transfer boundary layer resistance proximal to the cell surface—that is, we locally have an adsorption/desorption controlling system. Under these conditions, the extracellular concentration of species A along the cell surface is independent of any spatial mass transfer dependencies and is given by a prescribed function of time related to the physiological kinetics of the signal molecule delivery to the region—that is,

$$c_A|_{s=0} = c_A|_{s=0}(t)$$

For example, we consider here the following periodic "dosage" behavior of the extracellular signal molecule:

$$c_A|_{s=0}(t) = C_o[1 + \cos(bt)]$$

where C_o and b are known constants from the extracellular pathway analysis.

Now it can be seen that the mass balance equation given above is a first-order, linear differential equation with time-dependent coefficients, and it readily can be solved for any prescribed extracellular species concentration behavior. In the simplest mathematical case, where $c_A|_{s=0}(t) = C_o$, a constant, the mass balance equation is easily solved to yield

$$\underline{c}_{A(1)} = \underline{c}_{A(1)}(t=\infty)\,[1 - \exp[-Sa(k_d + C_o k_a)t]]$$

where we have added the subscript (1) to indicate the path number for reference here, and $\underline{c}_{A(1)}(t=\infty)$ is the steady-state solution obtained in Chapter 4:

$$\underline{c}_{A(1)}(t=\infty) = \frac{K\underline{c}_A^{\max} C_o}{1 + KC_o}$$

Using some typical values in cellular systems given in Table 5-1, this simple exponential build-up to the steady-state species

Table 5-1 Input Parameters for 4-Path Model

Path 1
$k_a = 1.0e05\ \mathrm{M^{-1}\ s^{-1}}$
$k_d = 3.0e\text{-}03\ \mathrm{s^{-1}}$
$\underline{c}_A^{max} = 1.0e04$
$aS = 1$
$C_o = 1.0e\text{-}08$ M
$b = 1/3600\ \mathrm{s^{-1}}$
Path 2
$k_f = 1.0e\text{-}02\ \mathrm{s^{-1}}$, $k_r = 1.0e\text{-}03\ \mathrm{s^{-1}}$
Path 3
$D_{BC} = 1.0e\text{-}11\ \mathrm{m^2/s}$
$R = 5.0e\text{-}05$ m
$c_{Bo} = 1.0e\text{-}06$ M
Path 4
$K_M = 1.0e\text{-}03$ M
$k_e\, c_{ET} = 5.0e\text{-}09\ \mathrm{M\ s^{-1}}$

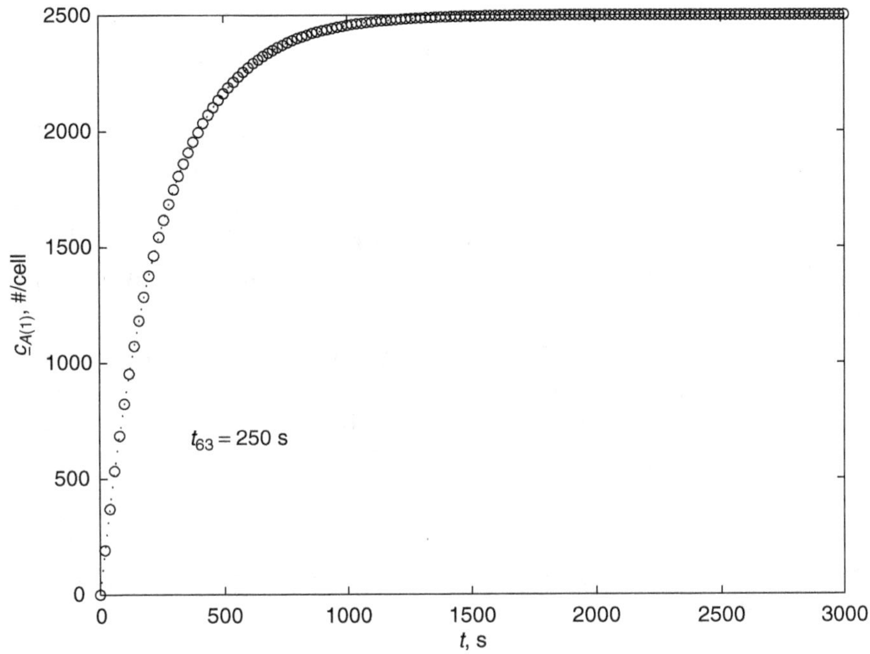

FIGURE 5-4 Simple exponential time-dependent buildup to the Langmuir adsorption behavior. The extracellular bulk signal molecule concentration is fixed in this calculation. Input parameters are given in Table 5-1.

surface amount is shown in Figure 5-4. From the time-dependent solution, we easily can see that the surface amount reaches 63% of its steady-state value in a time

$$t_{63} = [Sa(k_d + C_o k_a)]^{-1}$$

Now, for the oscillating or periodic extracellular concentration prescription just given, the mass balance equation can again readily be solved in closed-form to obtain

$$\underline{c}_{A(1)} = e^{-q(t)} \int_0^t Sak_a \underline{c}_A^{\max}[1 + C_o \cos(bt')]e^{q(t')}dt'$$

where

$$q(t) = Sa\left(k_d t + k_a C_o \left[t + \frac{1}{b}\sin(bt)\right]\right)$$

Using the values given in Table 5-1 and a characteristic oscillatory period on the order of an hour, an example solution is sketched in Figure 5-5.

Now consider the second pathway, which is a simple signal transduction process whereby the binding of species A on a cell surface receptor molecule leads to the release of a receptor-associated molecule B, as depicted in Figure 5-3. This would be a highly simplified view of a second messenger system for the living cell.

The dynamics of the release of the receptor-associated molecule B must be determined through a molecular-based study of the conformational changes in the membrane macromolecule upon the binding of molecule A. As noted in Chapter 4 in the discussion of conformational changes in macromolecules, this is an area of much needed quantitative research. However, ensemble averaging of molecular-based analyses eventually will

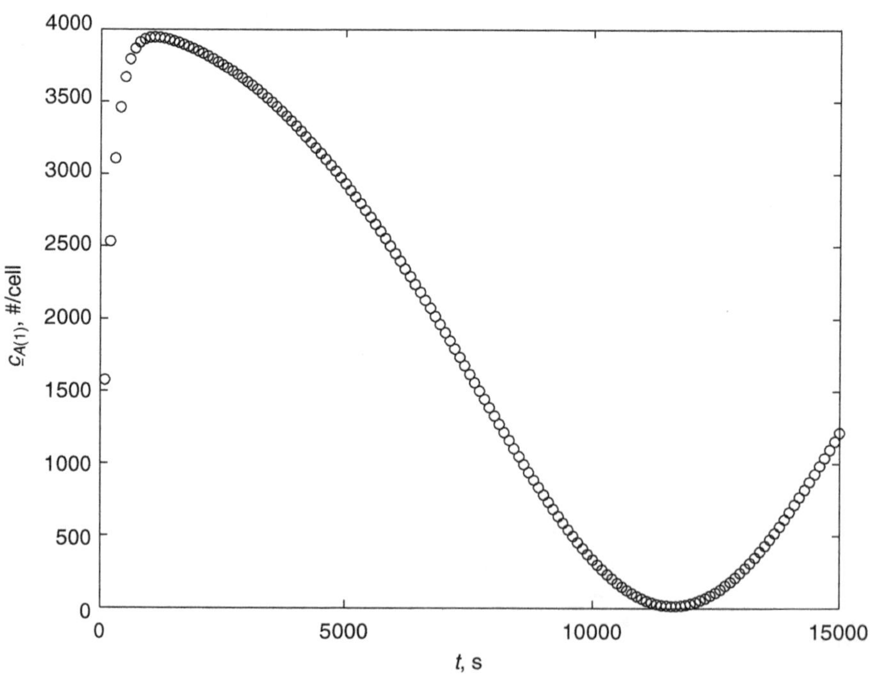

FIGURE 5-5 Ligand adsorption behavior for a time-dependent bulk signal molecule concentration. Input parameters are given in Table 5-1.

result in the casting of the behavior in terms of a macroscopic release rate constant, which we can speculate here for the purposes of illustration as

$$\frac{d\underline{c}_{B(2)}}{dt} = k_n^f \underline{c}_{A(1)}^n - k_m^r \underline{c}_{B(2)}^m$$

where k_n^f is an nth order forward rate constant for the release of species B, $\underline{c}_{B(2)}$ is the local amount of species B released along the internal surface of the cell, and k_m^r is an mth order reverse rate constant, where we have assumed here, for the sake of simplicity, that species B regulates its own release. All rate constants are determined from molecular simulation or some independent experimental analysis. In the linear case with $m = n = 1$, this equation admits a simple analytical solution subject to the particular expression for $\underline{c}_A(t)$, as determined from the first pathway analysis:

$$\underline{c}_{B(2)} = e^{-k_1^r t} \int_0^t k_1^f \underline{c}_{A(1)}(t') e^{k_1^r t'} dt'$$

where we have assumed that there is no "free" species B in the cell initially. Note that particular stoichiometric coefficients related to the release of B by the binding of A, as reflected by higher-ordered rate constants, would represent an example of signal amplification along the pathway.

Now once B has been released, we assume in this example that it undergoes a specific transport pathway to the particular cell region, or regions, containing the enzyme for the reaction step. This third transport pathway is described by the macroscopic species transport equation[5]

$$\frac{\partial c_B}{\partial t} = -\frac{\partial}{\partial \mathbf{r}} \cdot \mathbf{N}_B$$

where $\mathbf{N}_B$ is the flux of B by various possible transport mechanisms, such as simple Fickian diffusion, diffusiophoresis, and electrophoresis. Once the appropriate flux behavior is known, the macroscopic species balance equation can be solved subject

[5] See, e.g., Bird et al. (2002).

to the initial value at the beginning of the transport pathway to obtain the value at the end of the transport pathway where the enzymatic reaction can proceed.

Let's consider a specific example where an analytical solution can be developed for the purposes of illustration. We approximate the cell as a spherical body of radius R. The concentration of B at the cell surface, denoted by c_{Bo}, represents the free concentration of B as it is released from the membrane molecular complex and before it undergoes diffusion or transport. This free surface concentration is determined from the amount of B released from the second pathway expression just given divided by the volume occupied by the membrane complex molecules. The volume is approximately the plasma membrane thickness, δl times the external surface area of the cell, $4\pi R^2$, thus

$$c_{Bo}(t) = \frac{\underline{c}_{B(2)}(t)}{\underline{N}\delta l 4\pi R^2}$$

where $\underline{N}$ is Avogadro's number. Now, we will consider the mechanism of transport to be purely Fickian diffusion of B in the cell cytoplasm with an effective diffusivity D_{BC}. The mass transport equation in spherical coordinates and under spherically isotropic conditions is given by

$$\frac{\partial c_B}{\partial t} = D_{BC}\left[\frac{1}{r^2}\frac{\partial}{\partial r}\left(r^2\frac{\partial c_B}{\partial r}\right)\right]$$

where c_B is the concentration of B along the transport pathway, r is the radial coordinate, t is time, and D_{BC} is the diffusion coefficient of B in the cytoplasm. Note that transport equations also have a firm molecular-based foundation,[6] and they are also based on an ensemble averaging process of statistical mechanics. This foundation allows, for example, the determination of the diffusion coefficient from information about the particular molecular structure and interactions.

Now, we assume a zero initial concentration of B in the cytoplasm, and we can analytically integrate this mass transport,

[6] See, e.g., McQuarrie (1976), or Peters (2005).

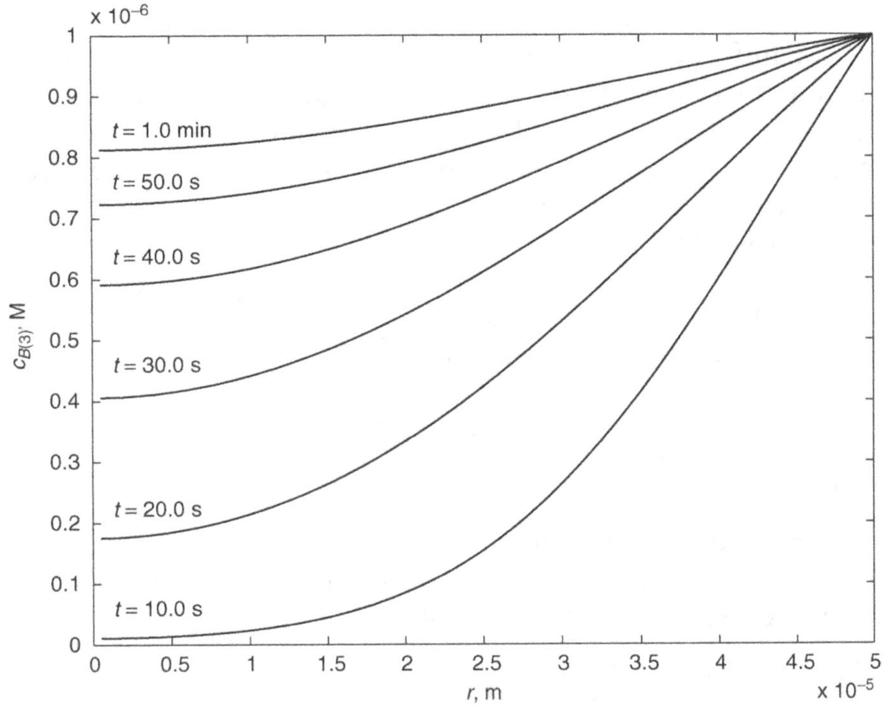

FIGURE 5-6 Time-dependent intracellular diffusion of a secondary signal molecule from a surface source. Input parameters are given in Table 5-1.

partial differential equation for any time-dependent surface condition

$$c_{Bo} = c_{Bo}(t)$$

to obtain[7]

$$c_B = \frac{2}{Rr}\Sigma_{n=1}^{\infty}\left[\exp\left(-\frac{D_{BC}n^2\pi^2 t}{R^2}\right)\sin\left(\frac{n\pi r}{R}\right)\right] \times \left(-n\pi\, D_{BC}(-1)^n \int_0^t \exp\left(\frac{D_{BC}n^2\pi^2 t'}{R^2}\right) c_{Bo}(t')dt'\right)$$

An example solution for a constant surface source is given in Figure 5-6.

[7] Carslaw and Jaeger (1959), Sec. 9.3.

Now, we consider that the enzyme involved in the final step here is uniformly distributed throughout the cell, such as in the mitochondria organelles. Then, we require the average concentration in the cell as a function of time, denoted as $c_{B(3)}(t)$. This readily is obtained by the integration of the previous expression over the entire spherical volume and then dividing by the volume to obtain

$$c_{B(3)}(t) = \frac{3}{R^3}\int_0^a c_B r^2 dr = \frac{6D_{BC}}{R^2}\Sigma_{n=1}^{\infty}\left[\exp\left(-\frac{D_{BC}n^2\pi^2 t}{R^2}\right)\right]$$

$$\times(-1)^{2n}\int_0^t \exp\left(\frac{D_{BC}n^2\pi^2 t'}{R^2}\right) c_{Bo}(t')dt'$$

This analytical expression quantitatively describes the transport pathway limitation to the process, and the result depends in a complicated manner on the surface concentration in this particular pathway, $c_{Bo}(t)$.

Finally, the enzymatic pathway dynamics will depend on the particular kinetics of the reaction. For simplicity, consider a Michaelis-Menten type enzymatic reaction scheme. The rate of production of the product C in the local region is denoted as $c_{C(4)}$ and is given by[8]

$$\frac{dc_{C(4)}}{dt} = \frac{k_e c_{ET} c_{B(3)}(t)}{K_M + c_{B(3)}(t)}$$

where k_e, K_M are Michaelis-Menten constants, c_{ET} is the constant total enzyme concentration in the local region, and $c_{B(3)}(t)$ is the concentration at the end of the transport pathway as determined in the previous transport pathway analysis problem. Note that the total enzyme concentration may have explicit time dependencies associated with its particular pathway

[8] See, e.g., Voet et al. (2006).

dynamics—for example, with its transcription process or some other interconnecting pathway. In any event, the previous first-order linear differential equation is easily integrated for the specified time-dependent variables to obtain the desired final result, in this example, for the rate of production of the enzymatic compound $c_{C(4)}$ as a function of time.

Although this analysis was highly simplified, and as noted earlier there may be hundreds of pathways involved in any particular cellular process, it does illustrate the essential features of dynamic analysis, such as nonlinearities, time lag, controlling pathways, microscopic to macroscopic descriptions, signal amplification, and so on. In an actual cellular process problem, each pathway associated with a particular cellular subsystem must be described in the integrated fashion as illustrated in our example. At the macroscopic level, as we have shown here, the limitations to the development of an overall analysis of cellular processes have much more to do with obtaining an accurate underlying molecularly based description than with any computational limitations, as we summarize below.

The Cell as a Dynamic Chemical Processing Plant

We have illustrated the paradigm by which overall cellular behavior may be quantified and described in real time. The importance of a real-time analysis has been emphasized throughout this book. The dynamic nature of cellular processes is critical to their function. The time orchestration of events is necessary for the cell to carry out its multitude of functions on a continuous basis. A disruption in the time sequencing of events can lead to critical breakdowns. For example, a simple, aberrant transport limitation can impede an entire cellular process in which it participates.

Indeed, we may view the cell as a highly complex and mobile chemical processing plant (Figure 5-7). It can readily change its phenotype to produce entirely new compounds and exhibit new functions. The cell has sophisticated safety and information

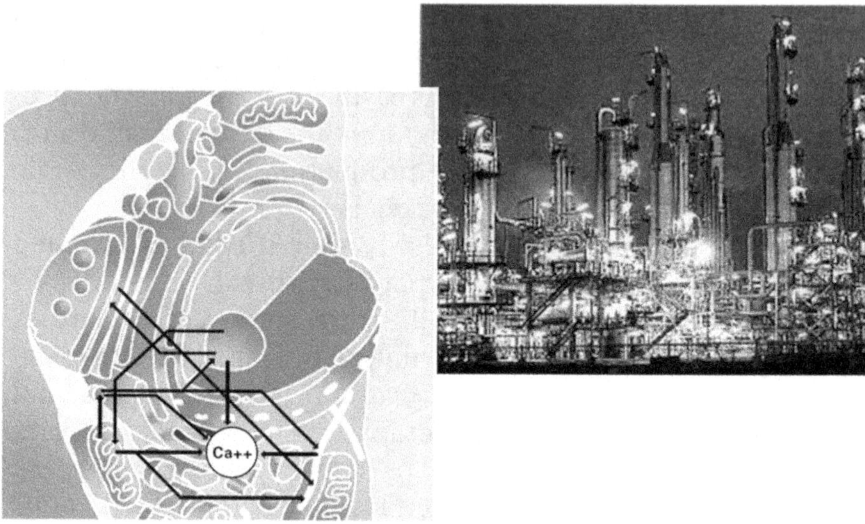

FIGURE 5-7 The living cell as a complex, dynamic biochemical production facility.

systems and can duplicate itself under the proper conditions. Interconnecting pathways allow for amplification, modulation, and control of cellular processes. Each pathway requires a sophisticated molecularly based analysis to connect pathway macroscopic dynamic relations to the underlying microscopic, or molecular, world. The goal of future research needs to be directed to the development of accurate qualitative and quantitative descriptions of dynamic pathways involved in cellular processes and to utilize advanced computing methods to connect these in an overall dynamic simulation of cellular processes.

References

Bird, R.B., Stewart, W.E., and Lightfoot, E.N. *Transport Phenomena,* Wiley, New York (2002).

Carslaw, H.S., and Jaeger, J.C., *Conduction of Heat in Solids,* 2d ed., Oxford University Press New York (1959).

Heinrich, R., and Schuster, S., *The Regulation of Cellular Systems,* Springer, New York (1996).

McQuarrie, D.A., *Statistical Mechanics,* Harper & Row, New York (1976).

Peters, M.H., *Molecular Thermodynamics and Transport Phenomena. Complexities of Scale in Space and Time,* McGraw-Hill Professional, New York (2005).

Pollack, G.H., *Cells, Gels, and the Engines of Life,* Ebner and Sons, New York (2001).

Sadler, T.W., *Langman's Medical Embryology,* Lippincott, Williams, & Wilkin, Baltimore (2004).

Voet, D., Voet, J.G., and Pratt, C.W., *Fundamentals of Biochemistry,* 2d ed., Wiley, New York (2006).

Appendix A

The Euler Angles

A description of the orientation of an object relative to a space-fixed, laboratory reference frame is not an easy task. Consider the simple example of describing the orientation of a rectangular block as shown in Figure A-1. We could define rotations of the block about a space-fixed Cartesian coordinate system as ϕ_x, ϕ_y, and ϕ_z, in order to obtain a particular orientation. Unfortunately, it is easy to show that if we rotate the block through specified values of ϕ_x, ϕ_y, and ϕ_z, the final orientation of the block depends on the order in which we perform these rotations.[1]

Fortunately, the problem of assigning rotational angles in unambiguous ways has been mathematically resolved. There are several different sets of rotational angle representations, and we consider only one of these here—namely, the so-called Euler angles. Other representations can be found in the references at the end of this appendix.

In order to succinctly see how the Euler angles are defined and used, first consider the simpler *two-dimensional* system shown in Figure A-2. The reference frame $(\bar{x}, \bar{y})$ is obtained by rotating

[1] Try this with a book as an example. Begin with the book lying flat on a table and define a space-fixed Cartesian coordinate system. Rotate the book around each axis by 90° in the order (ϕ_x, ϕ_y, and ϕ_z); then repeat the example by rotating in the order (ϕ_x, ϕ_z, and ϕ_y). Note how dramatically different the final orientations are in the two cases!

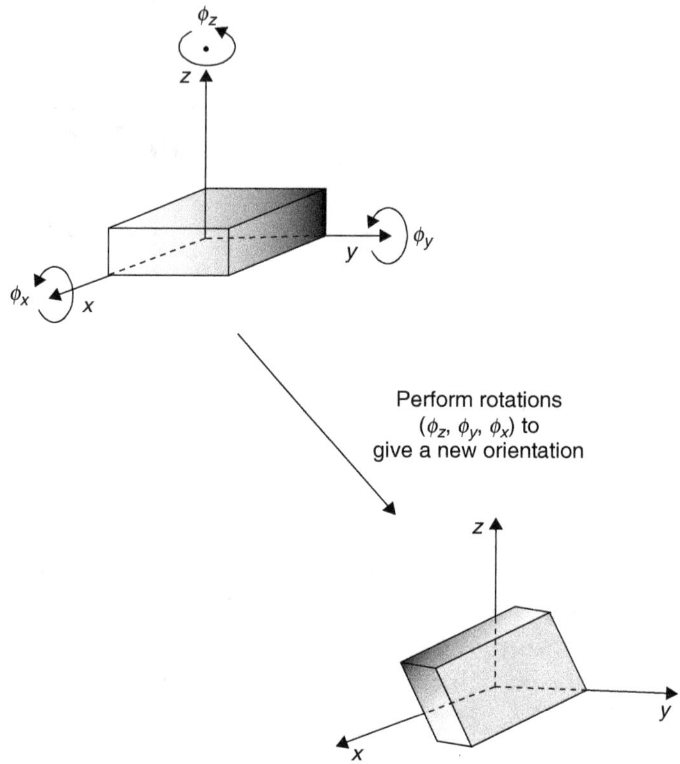

FIGURE A-1 Rotation of a block around Cartesian coordinates.

the (x, y) frame counterclockwise through an angle θ. We wish to describe a fixed vector **A** in both frames. We have

$$\begin{aligned} \mathbf{A} &= A_x\mathbf{e}_x + A_y\mathbf{e}_y \quad , \quad (x, y) \text{ frame} \\ &= A_{\bar{x}}\mathbf{e}_{\bar{x}} + A_{\bar{y}}\mathbf{e}_{\bar{y}} \quad , \quad (\bar{x}, \bar{y}) \text{ frame} \end{aligned} \tag{A-1}$$

where A_x and A_y are the components of **A** in the (x, y) frame, and $A_{\bar{x}}$ and $A_{\bar{y}}$ are the components of **A** in the $(\bar{x}, \bar{y})$ frame. Now, we can easily relate these two sets of components as follows:

$$\begin{aligned} A_x = \mathbf{e}_x \cdot \mathbf{A} &= A_{\bar{x}}(\mathbf{e}_x \cdot \mathbf{e}_{\bar{x}}) + A_{\bar{y}}(\mathbf{e}_x \cdot \mathbf{e}_{\bar{y}}) \\ &= A_{\bar{x}} \cos\theta - A_{\bar{y}} \sin\theta \end{aligned} \tag{A-2}$$

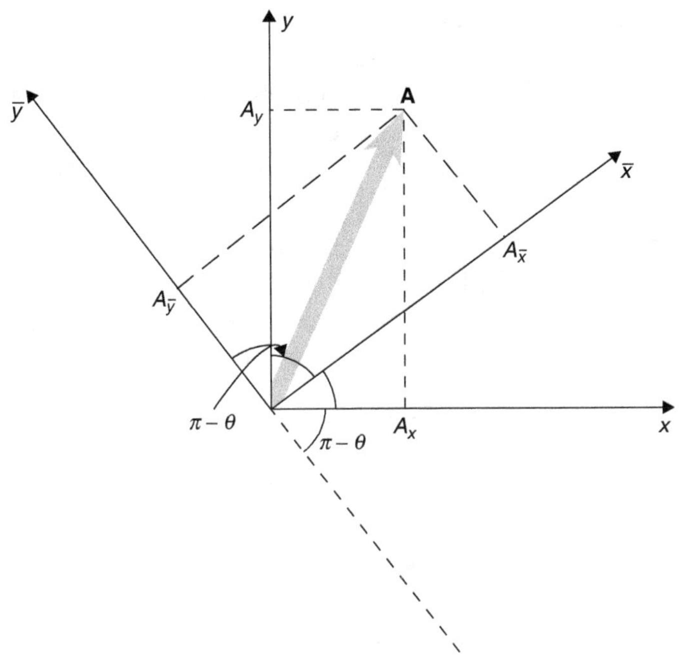

FIGURE A-2 Components of a vector A in two reference frames rotated by an angle θ from each other.

and

$$\begin{aligned} A_y = \mathbf{e}_y \cdot \mathbf{A} &= A_{\bar{x}}(\mathbf{e}_y \cdot \mathbf{e}_{\bar{x}}) + A_{\bar{y}}(\mathbf{e}_y \cdot \mathbf{e}_{\bar{y}}) \\ &= A_{\bar{x}} \sin\theta + A_{\bar{y}} \cos\theta \end{aligned} \tag{A-3}$$

In matrix notation

$$\begin{bmatrix} A_x \\ A_y \end{bmatrix} = \begin{bmatrix} \cos\theta & -\sin\theta \\ \sin\theta & \cos\theta \end{bmatrix} \begin{bmatrix} A_{\bar{x}} \\ A_{\bar{y}} \end{bmatrix} \tag{A-4}$$

or

$$A_i = \sum_{\bar{j}} e_{i\bar{j}} A_{\bar{j}} \tag{A-5}$$

where $i = x, y$ and the sum runs over all $\bar{j}$, $\bar{j} = \bar{x}, \bar{y}$, and $e_{i\bar{j}}$ are the matrix elements of the transformation matrix:

$$\left[e_{i\bar{j}} \right] = \begin{bmatrix} \cos\theta & -\sin\theta \\ \sin\theta & \cos\theta \end{bmatrix} \tag{A-6}$$

Now, consider a *three-dimensional* system. We wish to describe a fixed vector **A** in the (x, y, z) system that will be thrice rotated to a $(\bar{x}, \bar{y}, \bar{z})$ system. For ease of notation, we label the (x, y, z) axes as $(1, 2, 3)$ and the $(\bar{x}, \bar{y}, \bar{z})$ axes as $(\bar{1}, \bar{2}, \bar{3})$. All sets of axes are assumed to represent orthogonal coordinate systems.

The first rotation of axes is to rotate the (1, 2, 3) axes by an angle ϕ in the 1-2 plane, as shown in Figure A-3a. We label these new axes as $(\hat{1}, \hat{2}, \hat{3})$ and the transformation matrix, by analogy with the two-dimensional system, must be

$$\left[e_{i\hat{l}}\right] = \begin{bmatrix} \cos\phi & -\sin\phi & 0 \\ \sin\phi & \cos\phi & 0 \\ 0 & 0 & 1 \end{bmatrix} \tag{A-7}$$

or, in terms of vector components,

$$A_i = \sum_{\hat{l}} e_{i\hat{l}} A_{\hat{l}} \quad , \quad i = 1, 2, 3; \quad \hat{l} = \hat{1}, \hat{2}, \hat{3} \tag{A-8}$$

The second rotation, shown in Figure A-3b, rotates the $(\hat{1}, \hat{2}, \hat{3})$ axes by an angle θ in the $\hat{2}$-$\hat{3}$ plane. We label these new axes as $(\tilde{1}, \tilde{2}, \tilde{3})$ and the transformation of the matrix is

$$\left[e_{\hat{i}\tilde{k}}\right] = \begin{bmatrix} 1 & 0 & 0 \\ 0 & \cos\theta & -\sin\theta \\ 0 & \sin\theta & \cos\theta \end{bmatrix} \tag{A-9}$$

or, in terms of vector components,

$$A_{\hat{i}} = \sum_{\tilde{k}} e_{\hat{i}\tilde{k}} A_{\tilde{k}} \tag{A-10}$$

We note that by combining Eqs. (A-10) and (A-8), we have

$$\begin{aligned} A_i &= \sum_{\hat{l}} e_{i\hat{l}} \left[\sum_{\tilde{k}} e_{\hat{l}\tilde{k}} A_{\tilde{k}}\right] \\ &= \sum_{\hat{l}} \sum_{\tilde{k}} e_{i\hat{l}} e_{\hat{l}\tilde{k}} A_{\tilde{k}} \end{aligned} \tag{A-11}$$

or,

$$A_i = \sum_{\tilde{k}} e_{i\tilde{k}} A_{\tilde{k}} \tag{A-12}$$

where

$$e_{i\tilde{k}} = \sum_{\hat{l}} e_{i\hat{l}} e_{\hat{l}\tilde{k}} \tag{A-13}$$

Equation (A-12) gives the vector components in the (1, 2, 3) system terms of the vector components in the $(\tilde{1}, \tilde{2}, \tilde{3})$ system.

For our third and final rotation, we rotate the $(\tilde{1}, \tilde{2}, \tilde{3})$ system by an angle ψ in the $\tilde{1}$-$\tilde{2}$ plane, as shown in Figure A-3c.

We denote the final set of coordinates obtained in this manner as $(\bar{1}, \bar{2}, \bar{3})$, and the transformation matrix is given by

$$\left[e_{\tilde{k}\bar{j}}\right] = \begin{bmatrix} \cos\psi & -\sin\psi & 0 \\ \sin\psi & \cos\psi & 0 \\ 0 & 0 & 1 \end{bmatrix} \tag{A-14}$$

or, in terms of vector components,

$$A_{\tilde{k}} = \sum_{\bar{j}} e_{\tilde{k}\bar{j}} A_{\bar{j}} \tag{A-15}$$

By combining Eqs. (A-15) and (A-12), we obtain

$$A_i = \sum_{\bar{j}} e_{i\bar{j}} A_{\bar{j}} \tag{A-16}$$

where

$$e_{i\bar{j}} = \sum_{\tilde{k}} e_{i\tilde{k}} e_{\tilde{k}\bar{j}}, \tag{A-17}$$

or carrying out these operations using Eqs. (A-7), (A-9) and (A-14) gives

$$\left[e_{i\bar{j}}\right] = \begin{bmatrix} \cos\phi\cos\psi - \sin\phi\cos\theta\sin\psi & -\cos\phi\sin\psi - \sin\phi\cos\theta\cos\psi & \sin\phi\sin\theta \\ \sin\phi\cos\psi + \cos\phi\cos\theta\sin\psi & -\sin\phi\sin\psi + \cos\phi\cos\theta\cos\psi & -\cos\phi\sin\theta \\ \sin\theta\sin\psi & \sin\theta\cos\psi & \cos\theta \end{bmatrix} \tag{A-18}$$

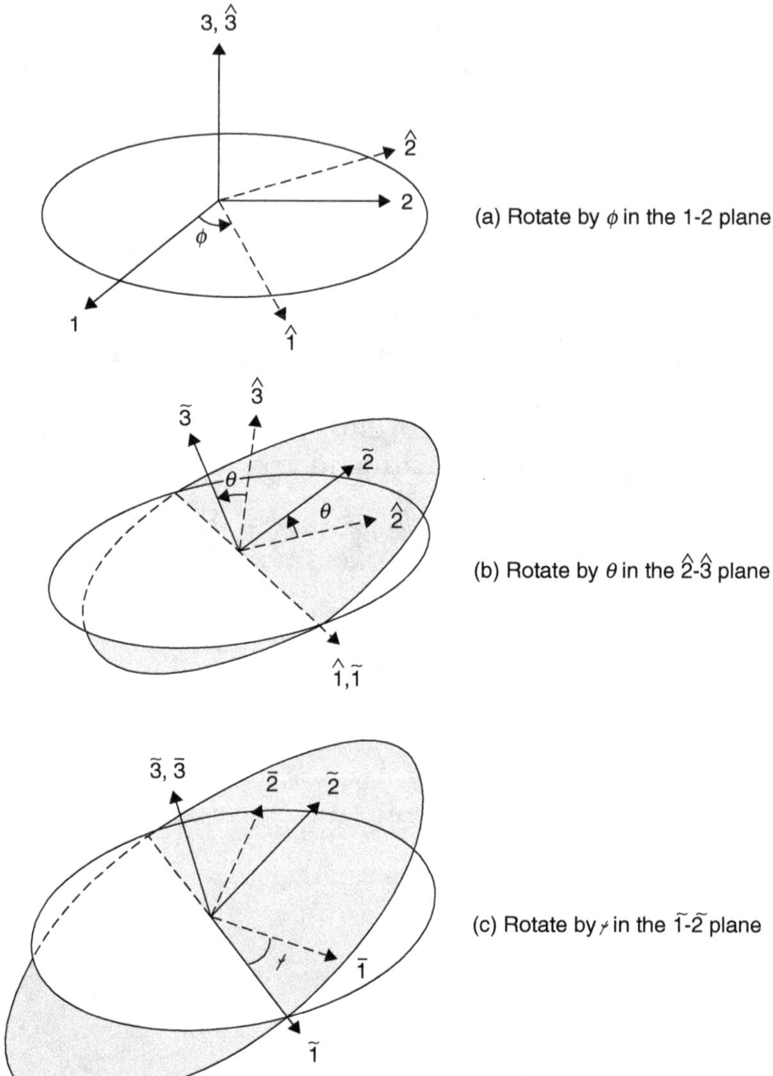

FIGURE A-3 The specific rotation of axes that define the Euler angles.

Equations (A-16) and (A-18) are used extensively in applications. The $(\bar{1}, \bar{2}, \bar{3})$ or $(\bar{x}, \bar{y}, \bar{z})$ system is taken to be fixed in the body of the molecule, and the (1, 2, 3) or (x, y, z) system is taken as the laboratory or space-fixed frame of reference. Then Eqs. (A-16) and (A-18) provide the prescription for computing the lab-frame components of a vector from knowledge of the body-frame components.

Appendix B

Liouville Equation, Fokker-Planck Equation (FP Brownian Dynamics), and Smoluchowski Equation (Sm Brownian Dynamics)

The Liouville Equation

The Liouville equation provides the complete molecular description of the system dynamics (biological molecules, solvent, external and molecular interaction forces, etc.) in terms of a probability density function for finding all of the system molecules with particular positions and velocities. More rigorously, the Liouville equation is developed in terms of so-called generalized coordinates and conjugate momenta. The generalized coordinates are the totality of the degrees of freedom in the entire system.

We denote the entirety of generalized coordinates as $q = \{q_1, q_2, \ldots, q_s\}$, the associated conjugate momenta as $p = \{p_1, p_2, \ldots, p_s\}$, and a probability density function f in this multidimensional space (phase space) as

$$f(q, p, t)dqdp$$

which represents the number of occurrences, or phase points, of the system between (q, p) and $(q + dq, p + dp)$ at any time t. The total number of phase points must satisfy a simple conservation relation

$$\frac{\partial f}{\partial t} = -\sum_{i=1}^{s}\left[\frac{\partial(f\dot{q}_i)}{\partial q_i} + \frac{\partial(f\dot{p}_i)}{\partial p_i}\right] \tag{B-1}$$

where s is the total number of degrees of freedom.

In working with generalized coordinates and conjugate momenta, Hamilton's equations of motion must be used[1]

$$\dot{q}_i = \frac{\partial H}{\partial p_i} \quad \text{and} \quad \dot{p}_i = -\frac{\partial H}{\partial q_i} \tag{B-2}$$

Hamilton's equations of motion constitute $2s$ first-order differential equations for the $2s$ unknown functions $p_i(t)$ and $q_i(t)$.

Substituting Hamilton's equations of motion into our conservation equation and simplifying gives

$$\frac{\partial f}{\partial t} = -\sum_{i=1}^{s}\left[\frac{\partial H}{\partial p_i}\frac{\partial f}{\partial q_i} - \frac{\partial H}{\partial q_i}\frac{\partial f}{\partial p_i}\right] \tag{B-3}$$

which is the desired Liouville equation written in terms of the generalized coordinates and conjugate momenta.

Now, we can show an important property of the generalized Liouville equation. If we write

$$f = f(q, p, t) \tag{B-4}$$

according to the chain rule for partial derivatives

$$\frac{df}{dt} = \frac{\partial f}{\partial t} + \sum_{i=1}^{s}\left[\frac{\partial f}{\partial q_i}\dot{q}_i + \frac{\partial f}{\partial p_i}\dot{p}_i\right] \tag{B-5}$$

[1] See, e.g., Peters (2005).

Comparing Eqs. (B-3) and (B-5) yields

$$\frac{df}{dt} = 0 \tag{B-6}$$

Thus, the density of phase points behaves as an "incompressible fluid." In other words, a differential volume element in phase space containing a certain number of phase points can change its shape, but not its size.

Now, consider the generalized coordinates necessary for the description of a rigid macromolecule in a host solvent, including the possibility of a nearby macromolecule or surface, as depicted in Figure B-1 and given explicitly as

$$(q, p) \equiv \left(\mathbf{r}^N, \mathbf{p}^N, \mathbf{R}, \Omega, \mathbf{P}, \mathbf{P}_\Omega\right) \tag{B-7}$$

Here $(\mathbf{r}^N, \mathbf{p}^N)$ refer to the position and momentum set of solvent molecules $(\mathbf{r}_1, \mathbf{r}_2, \ldots, \mathbf{r}_N, \mathbf{p}_1, \mathbf{p}_2, \ldots, \mathbf{p}_N)$. Also, Ω and $\mathbf{P}_\Omega$ are shorthand notation for the set of Euler angles (ϕ, ψ, θ) and associated conjugate momenta $(P_\phi, P_\psi, P_\theta)$ for the macromolecule or B (Brownian) particle, respectively; it is to be understood that these are not vectors in the usual sense. Finally, $(\mathbf{R}, \mathbf{P})$ are the center of mass position and momentum of the B-particle.

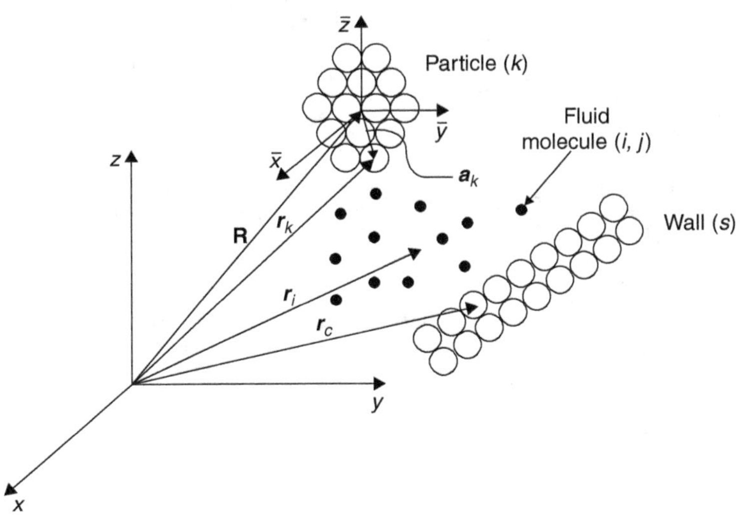

FIGURE B-1 Illustration of the variables used to describe the orientation and location of a structured macromolecule near another structured surface. Overbars indicate particle frame coordinates.

For our system, assuming pairwise additivity of the interaction forces, the Hamiltonian or total energy can be expressed as

$$H = T + \sum_{i=1}^{N} \sum_{\substack{j=1 \\ j<i}}^{N} u_{ff}(\mathbf{r}_i, \mathbf{r}_j) + \sum_{i=1}^{N} \sum_{k} u_{fp}(\mathbf{r}_i, \mathbf{R}, \Omega; \bar{\mathbf{a}}_k)$$

$$+ \sum_{i=1}^{N} \sum_{s} u_{fw}(|\mathbf{r}_i - \mathbf{r}_s|) + \sum_{k} \sum_{s} u_{p\omega}(\mathbf{r}_s, \mathbf{R}, \Omega; \bar{\mathbf{a}}_k) \quad \text{(B-8)}$$

where T is the total kinetic energy, and $u_{ff}(\mathbf{r}_i, \mathbf{r}_j)$, $u_{fp}(\mathbf{r}_i, \mathbf{R}, \Omega; \bar{\mathbf{a}}_k)$, $u_{p\omega}(\mathbf{r}_s, \mathbf{R}, \Omega; \bar{\mathbf{a}}_k)$ are the solvent-solvent, solvent-particle, particle-surface pair molecular interaction potentials, respectively. Also, $\mathbf{r}_i$ or $\mathbf{r}_j$ are location vectors for the fluid molecules, $\mathbf{r}_s$ is the location vector for a surface molecule, $\bar{\mathbf{a}}_k$ is the location vector of the kth constituent of the B-particle in a frame of reference fixed with respect to the B-particle and with origin at the center of mass $\mathbf{R}$ (see Figure B-1). The locator vectors $\bar{\mathbf{a}}_k$ are constants and fix the molecular structure of the B-particle; they are assumed known a priori. The total kinetic energy is given by the sum of the translational and rotational kinetic energy:

$$T = \sum_{i=1}^{N} \frac{1}{2} \frac{p_i^2}{m} + \frac{1}{2} \frac{P^2}{M} + T_{\text{rot}} \quad \text{(B-9)}$$

The rotational kinetic energy can be expressed in terms of the generalized momenta as[2]

$$T_{\text{rot}} = \frac{1}{2} \frac{1}{I_{\bar{1}} \sin^2 \theta} [\sin \psi (P_\phi - \cos \theta P_\psi) + \cos \psi \sin \theta P_\theta]^2$$

$$+ \frac{1}{2} \frac{1}{I_{\bar{2}} \sin^2 \theta} [\cos \psi (P_\phi - \cos \theta P_\psi) - \sin \psi \sin \theta P_\theta]^2$$

$$+ \frac{1}{2} \frac{1}{I_{\bar{3}}} P_\psi^2 \quad \text{(B-10)}$$

[2] Peters (1999).

where the principal moment $I_{\bar{i}}$ is the $\bar{i}^{\text{th}}$ component of the moment of inertia of the B-particle in the body-fixed frame. Note that the principal moments of inertia are constant for a rigid B-particle.

The functional dependencies shown for the interaction potentials u_{fp} and $u_{p\omega}$ can be seen as follows. We have that (Figure B-1)

$$u_{fp} = u_{fp}(|\mathbf{r}_i - (\mathbf{a}_k + \mathbf{R})|) \tag{B-11}$$

where $\mathbf{a}_k$ is the locator vector with respect to the laboratory frame. Relative to the B-particle frame, we have

$$\mathbf{a}_k = \mathbf{S} \cdot \bar{\mathbf{a}}_k \tag{B-12}$$

where $\mathbf{S}$ is the transformation matrix that relates the locator vector in the B-particle frame $\bar{\mathbf{a}}_k$ to the laboratory or inertial frame $\mathbf{a}_k$. $\mathbf{S}$ depends explicitly on the Euler angles—that is, in component notation

$$a_{k_i} = \sum_{\bar{j}} e_{i\bar{j}} \bar{a}_{k_{\bar{j}}} \tag{B-13}$$

and the transformation matrix is given by[3]

$$[e_{i\bar{j}}] = \begin{bmatrix} \cos\phi\cos\psi - \sin\phi\cos\theta\sin\psi & -\cos\phi\sin\psi - \sin\phi\cos\theta\cos\psi & \sin\phi\sin\theta \\ \sin\phi\cos\psi + \cos\phi\cos\theta\sin\psi & -\sin\phi\sin\psi + \cos\phi\cos\theta\cos\psi & -\cos\phi\sin\theta \\ \sin\theta\sin\psi & \sin\theta\cos\psi & \cos\theta \end{bmatrix} \tag{B-14}$$

Because the particle is assumed rigid, the locator vectors $\bar{\mathbf{a}}_k$ in the B-particle frame are constants. Thus, we can write that

$$u_{fp} = u_{fp}(\mathbf{r}_i, \mathbf{R}, \Omega; \bar{\mathbf{a}}_k) \tag{B-15}$$

A similar analysis applies to $u_{p\omega}$.

[3] See Appendix A.

Fokker-Planck Equation

To obtain the Fokker-Planck (FP) equation, a multiple time scale analysis of the Liouville equation is developed. First, the Liouville equation is nondimensionalized or scaled to obtain[4]

$$\frac{\partial f^*}{\partial t^*} = (-N_{\text{Kn}}^{-1} L_f^* + \gamma L_p^*) f^* \tag{B-16}$$

where $N_{\text{Kn}} = \frac{r_0}{R_0}$, $\gamma = m/M$, and L_f^* and L_p^* are particular nondimensional Liouville operators for this system, and the $*$ indicates a dimensionless variable. In the Knudsen number definition, r_0 is a characteristic length scale for the fluid intermolecular interaction force and, for an isolated Brownian particle, R_0 is a characteristic length scale for the interaction force between the macromolecule and the fluid, often taken as the effective overall macromolecular diameter. Also, $\gamma = m/M$ is a mass ratio term with M being the total mass of the macromolecule and m is the mass of a solvent molecule.

We also introduce the reduced or contracted distribution function for the B-particle only as

$$\Psi(\mathbf{R}, \mathbf{P}, \Omega, \mathbf{P}_\Omega, t) = \int_{\text{all } \mathbf{r}^N, \mathbf{p}^N} f\left(\mathbf{r}^N, \mathbf{p}^N, \mathbf{R}, \mathbf{P}, \Omega, \mathbf{P}_\Omega, t\right) d\mathbf{r}^N d\mathbf{p}^N \tag{B-17}$$

where $\Psi(\mathbf{R}, \mathbf{P}, \Omega, \mathbf{P}_\Omega, t)$ is the probability of finding the B-particle between $(\mathbf{R}, \mathbf{P}, \Omega, \mathbf{P}_\Omega)$ and $(\mathbf{R} + d\mathbf{R}, \mathbf{P} + d\mathbf{P}, \Omega + d\Omega, \mathbf{P}_\Omega + d\mathbf{P}_\Omega)$ at time t.

Now, the mass ratio is typically very small so that the distribution function can be expanded in terms of γ, and introducing multiple time scales $(\tau_0, \tau_1, \tau_2, \ldots)$, we can write

$$\begin{aligned} f &= f\left(\mathbf{r}^N, \mathbf{p}^N, \mathbf{R}, \mathbf{P}, \Omega, \mathbf{P}_\Omega, \tau_0, \tau_1, \tau_2, \ldots\right) \\ &= f^{(0)} + \gamma f^{(1)} + \gamma^2 f^{(2)} + \cdots \end{aligned} \tag{B-18}$$

The expansion analysis is conducted through second order in mass ratio and time scales,[5] leading to a specific expression,

[4] Peters (1999).
[5] Peters (1999).

or Fokker-Planck equation, for the total time behavior of the contracted distribution function for the B-particle, as

$$\frac{\partial \Psi}{\partial t} + \frac{\mathbf{P}}{M} \cdot \frac{\partial \Psi}{\partial \mathbf{R}} + \left[\langle \mathbf{F}_f \rangle_{\text{eq}} - \frac{\partial}{\partial \mathbf{R}} \left(\sum_k \sum_s u_{p\omega} \right) \right] \cdot \frac{\partial \Psi}{\partial \mathbf{P}}$$

$$+ \frac{\partial T_{\text{rot}}}{\partial \mathbf{P}_\Omega} \cdot \frac{\partial \Psi}{\partial \Omega} + \left[\langle \mathbf{T}_f \rangle_{\text{eq}} - \frac{\partial}{\partial \Omega} \left(T_{\text{rot}} + \sum_k \sum_s u_{p\omega} \right) \right] \cdot \frac{\partial \Psi}{\partial \mathbf{P}_\Omega}$$

$$= kT \left\{ \frac{\partial}{\partial \mathbf{P}} \cdot \left(\zeta_T \cdot \left[\Psi \frac{\mathbf{P}}{MkT} + \frac{\partial \Psi}{\partial \mathbf{P}} \right] + \zeta_{TR} \cdot \left[\Psi \frac{1}{kT} \frac{\partial T_{\text{rot}}}{\partial \mathbf{P}_\Omega} + \frac{\partial \Psi}{\partial \mathbf{P}_\Omega} \right] \right) \right.$$

$$\left. + \frac{\partial}{\partial \mathbf{P}_\Omega} \cdot \left(\zeta_R \cdot \left[\Psi \frac{1}{kT} \frac{\partial T_{\text{rot}}}{\partial \mathbf{P}_\Omega} + \frac{\partial \Psi}{\partial \mathbf{P}_\Omega} \right] + \zeta_{RT} \cdot \left[\Psi \frac{\mathbf{P}}{MkT} + \frac{\partial \Psi}{\partial \mathbf{P}} \right] \right) \right\} \tag{B-19}$$

where k is Boltzmann's constant, T is absolute temperature, and T_{rot} is the rotational kinetic energy of the B-particle given here.

In general, the force and torque acting on the B-particle by the fluid are defined, respectively, by

$$\mathbf{F}_f = -\frac{\partial}{\partial \mathbf{R}} \left[\sum_{i=1}^{N} \sum_k u_{fp} \right] \tag{B-20}$$

$$\mathbf{T}_f = -\frac{\partial}{\partial \Omega} \left[\sum_{i=1}^{N} \sum_k u_{fp} \right] \tag{B-21}$$

Further, $\langle \mathbf{F}_f \rangle_{\text{eq}}$ and $\langle \mathbf{T}_f \rangle_{\text{eq}}$ are the equilibrium average force and torque, respectively, acting on the B-particle by the fluid. For an isolated B-particle, these terms vanish, but they are finite when other B-particles or surfaces are present.

The translational, rotational, and coupled translational-rotational friction tensors in Eq. (B-19) are given by

$$\zeta_T \equiv \frac{1}{kT} \int_0^\infty \left[\langle \mathbf{F}_f(s) \mathbf{F}_f(0) \rangle_{\text{eq}} - \langle \mathbf{F}_f \rangle_{\text{eq}}^2 \right] ds \tag{B-22}$$

$$\zeta_{TR} \equiv \frac{1}{kT} \int_0^\infty \left[\langle \mathbf{F}_f(s) \mathbf{T}_f(0) \rangle_{\text{eq}} - \langle \mathbf{F}_f \rangle_{\text{eq}} \langle \mathbf{T}_f \rangle_{\text{eq}} \right] ds \tag{B-23}$$

$$\zeta_{RT} \equiv \frac{1}{kT}\int_0^\infty \left[\langle \mathbf{T}_f(s)\mathbf{F}_f(0)\rangle_{\text{eq}} - \langle \mathbf{T}_f\rangle_{\text{eq}}\langle \mathbf{F}\rangle_{\text{eq}}\right] ds \tag{B-24}$$

$$\zeta_R \equiv \frac{1}{kT}\int_0^\infty \left[\langle \mathbf{T}_f(s)\mathbf{T}_f(0)\rangle_{\text{eq}} - \langle \mathbf{T}_f\rangle^2_{\text{eq}}\right] ds \tag{B-25}$$

Note that the friction tensors, Eqs. (B-22) through (B-25), depend on the separation distance and orientation of the B-particle with respect to other B-particles or surfaces present. In continuum fluid mechanics, this latter effect is known as *hydrodynamic interaction.* In general, the friction tensors can be determined from these integrals through equilibrium canonical molecular dynamics for any given B-particle position and orientation in the system. This result is the heart of the multiple time scale analysis. The host fluid quickly relaxes to the potential field created by the B-particle in a period of picoseconds or less under the conditions of small Knudsen numbers. Mathematically, these integrals converge rapidly under these conditions. During this short-time period, there is no significant change in the B-particle position or orientation.

Now, the numerical or finite-difference solution to the FP equation leads to a Brownian dynamics algorithm, or long-time, BD time step, for determining the macromolecular displacement.[6] The friction tensors, however, must be determined at the start of each FP BD time step via MD as just explained. The MD step requires the (local) equilibrium hydration of the macromolecule to begin. Once the initial B-particle state and friction tensors are specified, the B-particle is moved according to the FP equation over one BD time step (typically on the order of nanoseconds). Then, the entire process must be repeated for the next FP BD time step—that is, hydration and MD followed by BD. However, for an isolated B-particle, the friction tensor is invariant in the body-frame of the particle and need be determined only once. Before giving additional specific expressions for carrying out actual numerical solutions, a further simplification often exists for biological macromolecules.

[6] See Peters (1999).

This simplification leads to the so-called Smoluchowski (Sm) equation to describe the long-time behavior rather than the FP equation. The Smoluchowski equation does not require the B-particle linear and angular momenta as independent variables and is, therefore, somewhat simpler to use. Next, we concentrate on the development of the Sm equation and its associated Brownian dynamics method.

The Smoluchowski Equation

Using operator notation and the particle angular momentum **L** it is convenient to reexpress the FP equation in more compact form as[7]

$$\frac{\mathcal{D}\Psi}{\mathcal{D}t} = kT\,[\mathbf{G}_p \quad \mathbf{G}_L]\begin{bmatrix}\zeta_T & \zeta_{TR}\\ \zeta_{RT} & \zeta_R\end{bmatrix}\begin{bmatrix}Q_p\Psi\\ Q_L\Psi\end{bmatrix} \tag{B-26}$$

Now, following Wilemski (1976), we derive the Sm equation from a regular perturbation expansion of the FP equation. This formality allows us to set definite limits on the applicability of the Sm equation, which is important in applications and in establishing higher-order corrections. First, we define the following dimensionless variables (denoted by asterisk symbols):

$$\Psi = \Psi^* \frac{n_0}{(R_0 MkT)^3} \tag{B-27}$$

$$\mathbf{P} = \mathbf{P}^*(MkT)^{1/2} \tag{B-28}$$

$$\mathbf{R} = \mathbf{R}^* R_0 \tag{B-29}$$

$$t = t^* t_0 = t^*[R_0/(kT/M)^{1/2}] \tag{B-30}$$

$$\mathbf{L} = \mathbf{L}^* R_0 (MkT)^{1/2} \tag{B-31}$$

$$\mathbf{F}' = \mathbf{F}'^*(kT/R_0) \tag{B-32}$$

$$\mathbf{T}' = \mathbf{T}'^*(kT) \tag{B-33}$$

[7] The operators in Eq. (B-26) are defined in Peters (2000).

$$\zeta_T = \zeta_T^* \zeta_0 \tag{B-34}$$

$$\zeta_{TR} = \zeta_{TR}^* R_0 \zeta_0 \tag{B-35}$$

$$\zeta_{RT} = \zeta_{RT}^* R_0 \zeta_0 \tag{B-36}$$

$$\zeta_R = \zeta_R^* R_0^2 \zeta_0 \tag{B-37}$$

Substituting the dimensionless variables into the FP equation, Eq.(B-26), yields

$$\frac{\mathcal{D}^* \Psi^*}{\mathcal{D}^* t^*} = \frac{1}{N_{\mathrm{St}}} \left\{ \begin{bmatrix} \mathbf{G}_p^* & \mathbf{G}_L^* \end{bmatrix} \begin{bmatrix} \zeta_T^* & \zeta_{TR}^* \\ \zeta_{RT}^* & \zeta_R^* \end{bmatrix} \begin{bmatrix} \mathcal{Q}_P^* & \Psi^* \\ \mathcal{Q}_L^* & \Psi^* \end{bmatrix} \right\} \tag{B-38}$$

where N_{St} is the particle Stokes number:

$$N_{\mathrm{St}} = \frac{\zeta_0^{-1} M}{R_0/(kT/M)^{1/2}} = \frac{t_{\mathrm{relax}}}{t_0} \tag{B-39}$$

The Stokes number is the ratio of the particle relaxation time $\zeta_0^{-1} M$ to the characteristic time constant t_0. The Sm equation represents a first-order expansion of the FP equation under the conditions

$$N_{\mathrm{St}} << 1 \tag{B-40}$$

or

$$R_0 >> \left(\frac{kT}{M\beta^2} \right)^{1/2} \tag{B-41}$$

where $\beta \equiv \zeta_0/M$. The inequality relationship, Eq. (B-41), was first given by Chandrasekhar (1943), and it is important in establishing limitations of the Sm equation. For example, relatively strong external forces (small characteristic R_0) could negate the applicability of the Sm equation.[8]

[8] More quantitatively, R_0 is the characteristic length over which the potential-associated with any (nonhydrodynamic) external force or torque acting on the particle changes by $O(kT)$. For electrically neutral systems, R_0 is usually on the order of the particle radius. Under these conditions, a particle with radius of 1 micron and density of 2.0 g/cm^3 in water at 20°C has a Stokes number of 3.1×10^{-4}. On the other hand, that same particle in an electrical field of 1000 V/cm carrying 1000 elementary charges gives an R_0 of 2.6×10^{-8} cm and a Stokes number of 1.2.

Now, we introduce the following definitions:

- configurational-orientational density function:

$$n(\mathbf{R}, \theta, \phi, \psi, t) \equiv \int\int \Psi d\mathbf{P} d\mathbf{L} \tag{B-42}$$

- translational flux ($\sim$ average particle translational velocity, $\langle \mathbf{P} \rangle / M$):

$$\mathbf{j}_R \equiv n\langle \mathbf{P}/M \rangle \equiv \int\int (\mathbf{P}/M) \Psi d\mathbf{P} d\mathbf{L} \tag{B-43}$$

- angular flux ($\sim$ average particle angular velocity, $\mathbf{I}^{-1} \cdot \langle \mathbf{L} \rangle$):

$$\mathbf{j}_w \equiv n\mathbf{I}^{-1} \cdot \langle \mathbf{L} \rangle \equiv \int\int \mathbf{I}^{-1} \cdot \mathbf{L} \Psi d\mathbf{P} d\mathbf{L} \tag{B-44}$$

Or, in dimensionless terms, these relationships are

$$n = n_0 n^*(\mathbf{R}^*, \theta, \phi, \psi, t^*) \tag{B-45}$$

$$\mathbf{j}_R = \mathbf{j}_R^* n_0 (kT/M)^{1/2} \tag{B-46}$$

$$\mathbf{j}_w = \mathbf{j}_w^* n_0 (kT/M)^{1/2} / R_0 \tag{B-47}$$

Now, by noting that the perturbation solution to Eq. (B-38) is "regular" in the limit $N_{St} \to 0$, the solution can be written in a Poincaré-type expansion as

$$\Psi = \sum_{n=0}^{\infty} \epsilon^n \Psi^{(n)}, \quad \epsilon \to 0+ \tag{B-48}$$

where the smallness parameter $\epsilon \equiv N_{St}$.

It is now straightforward to carry out the perturbative solution to the FP equation, Eq. (B-38) using the expansion, Eq. (B-48), along with the number density and flux relationships, Eqs. (B-42) through (B-44). Here we present only the final results, the *dimensionless* Smoluchowski equation, and leave the more tedious mathematical steps to the cited reference.[9] The

[9] Peters (2000). Also see Condiff and Dahler (1966), for the isolated B-particle case.

dimensionless Smoluchowski equation can be written to $0(\epsilon^2)$ as

$$\frac{\partial n}{\partial t} = \epsilon[\partial/\partial\mathbf{R}\ (\partial/\partial\alpha\cdot\Lambda)]\begin{bmatrix}\mathbf{D}_T & \mathbf{D}_{TR}\\ \mathbf{D}_{RT} & \mathbf{D}_R\end{bmatrix}\begin{bmatrix}\partial n/\partial\mathbf{R} - n\mathbf{F}'\\ (\partial/\partial\alpha\cdot\Lambda)n - n\mathbf{T}'\end{bmatrix} + 0(\epsilon^2) \qquad \text{(B-49)}$$

where the rotational operator $\frac{\partial}{\partial\alpha}\cdot\Lambda$ is defined as

$$\left(\frac{\partial}{\partial\alpha}\cdot\Lambda\right) \doteq \left[\csc\theta\frac{\partial}{\partial\theta}\sin\theta\ \ \frac{\partial}{\partial\phi}\ \ \frac{\partial}{\partial\psi}\right] \times \begin{bmatrix}\cos\phi & \sin\phi & 0\\ -\cot\theta\sin\phi & \cot\theta\cos\phi & 1\\ \csc\theta\sin\phi & -\csc\theta\cos\phi & 0\end{bmatrix} \qquad \text{(B-50)}$$

The x-component, for example, is

$$\csc\theta\frac{\partial}{\partial\theta}\sin\theta\cos\phi - \frac{\partial}{\partial\phi}\cot\theta\sin\phi + \frac{\partial}{\partial\psi}\csc\theta\sin\phi \qquad \text{(B-51)}$$

In the Sm equation, Eq. (B-49), the *dimensionless* particle diffusion tensors are given by

$$\mathbf{D}_T \equiv [\zeta_T - \zeta_{TR}\cdot\zeta_R^{-1}\cdot\zeta_{RT}]^{-1} \qquad \text{(B-52)}$$

$$\mathbf{D}_{TR} \equiv [\zeta_{RT} - \zeta_R\cdot\zeta_{TR}^{-1}\cdot\zeta_T]^{-1} \qquad \text{(B-53)}$$

$$\mathbf{D}_R \equiv [\zeta_R - \zeta_{RT}\cdot\zeta_T^{-1}\cdot\zeta_{TR}]^{-1} \qquad \text{(B-54)}$$

$$\mathbf{D}_{RT} \equiv [\zeta_{TR} - \zeta_T\cdot\zeta_{RT}^{-1}\cdot\zeta_R]^{-1} \qquad \text{(B-55)}$$

These so-called generalized Einstein relationships (in dimensionless terms here) are identical in form to those for an isolated Brownian particle. The subtle difference lies in the defining equations for the friction tensors given by Eqs. (B-22) through (B-25). These expressions include the presence of external surfaces, or other macromolecules, through both the nonzero equilibrium average force and torque and the time-dependent force and torque acting on the particle by the fluid molecules, which are simultaneously in the presence of the potential field of other particles or surfaces. Additionally, the nonzero equilibrium average force and torque directly contribute to the convective motions of the particle through the terms $\mathbf{F}'$ and $\mathbf{T}'$.

We note that the smallness parameter $\epsilon(=N_{\mathrm{St}})$ can be recast in terms of the dimensionless diffusion $D_0 \equiv kT/\zeta_0$ as

$$\epsilon = \frac{D_0}{R_0(kT/M)^{1/2}} \equiv \frac{1}{N_{\mathrm{Pe}}} \tag{B-56}$$

where N_{Pe} is a type of Peclet number, which is the ratio of the diffusion time scale (R_0^2/D_0) to the characteristic time scale (t_0) for particle translation or rotation due to the external force or torque field (or possibly an initial state). The asymptotic analysis previously outlined shows that the Peclet number must be large for the Smoluchowski equation to hold. For example, strong external force or torque fields lead to relatively small values of R_0, which may negate the use of the Smoluchowski equation. Another typical violation occurs for large values of M, the particle mass ("inertial effects"). Either of these situations may be remedied by the use of the more cumbersome Fokker-Planck equation or by deriving higher-order corrections to the Sm equation. The general solution to the Sm equation can be obtained by numerical solution that constitutes the BD method. Because of its importance in applications, we consider this solution in some detail here.

Sm Brownian Dynamics Method

Because of the inherent complexities in the molecular-based description of the Smoluchowski equation and the numerous potential applications to complex, practical systems, in this section we give a general numerical solution based on so-called Brownian dynamics method.

As shown previously,[10] a Brownian dynamics, numerical solution to the Smoluchowski equation can be obtained by deriving an analytical solution for small, finite time intervals. In doing this, we consider that at time $t=0$, the position and orientation

[10] Ermak and McCammon (1978).

of the Brownian particle are exactly known—that is,

$$n = \delta(\mathbf{R} - \mathbf{R}_0)\delta(\Omega - \Omega_0), \ \text{at } t = 0 \tag{B-57}$$

where δ is the Dirac delta function. Now, for small times $t\,(t > 0)$, we can assume that over the time interval $(0, t)$ all spatial and orientational functions are approximately constant at their initial values—that is,

$$\mathbf{F}' \cong \mathbf{F}'^0 \tag{B-58}$$

$$\mathbf{T}' \cong \mathbf{T}'^0 \tag{B-59}$$

$$\mathbf{D}_J \cong \mathbf{D}_J^0, \ J = T, TR, RT, R \tag{B-60}$$

and so on, where the superscript (0) indicates the known initial values. Also, with this approximation the rotational operator $(\partial/\partial\alpha \cdot \Lambda)$ is given by

$$\begin{aligned}(\partial/\partial\alpha \cdot \Lambda)^0 = \mathbf{e}_1 &\left[\cos\phi^0 \frac{\partial}{\partial\theta} - \cot\theta^0 \sin\phi^0 \frac{\partial}{\partial\phi} + \csc\theta^0 \sin\phi^0 \frac{\partial}{\partial\psi}\right] \\ +\mathbf{e}_2 &\left[\sin\phi^0 \frac{\partial}{\partial\theta} + \cot\theta^0 \cos\phi^0 \frac{\partial}{\partial\phi} - \csc\theta^0 \cos\phi^0 \frac{\partial}{\partial\psi}\right] \\ +\mathbf{e}_3 &\left[\frac{\partial}{\partial\phi}\right]\end{aligned} \tag{B-61}$$

Now we can readily show that the quantities in brackets in the previous equation are the derivative operators for rotations about the space-fixed Cartesian frame,[11]—that is,

$$\frac{\partial}{\partial\phi_x} = \cos\phi \frac{\partial}{\partial\theta} - \cot\theta \sin\phi \frac{\partial}{\partial\phi} + \csc\theta \sin\phi \frac{\partial}{\partial\psi} \tag{B-62}$$

$$\frac{\partial}{\partial\phi_y} = \sin\phi \frac{\partial}{\partial\theta} + \cot\theta \cos\phi \frac{\partial}{\partial\phi} - \csc\theta \cos\phi \frac{\partial}{\partial\psi} \tag{B-63}$$

$$\frac{\partial}{\partial\phi_z} = \frac{\partial}{\partial\phi} \tag{B-64}$$

[11] Peters (2000); also, see Goldstein (1980).

Thus, we can write

$$(\partial/\partial\alpha \cdot \Lambda)^0 = \frac{\partial}{\partial \Phi} \equiv \mathbf{e}_1 \frac{\partial}{\partial \phi_1} + \mathbf{e}_2 \frac{\partial}{\partial \phi_2} + \mathbf{e}_3 \frac{\partial}{\partial \phi_3} \qquad \text{(B-65)}$$

Using Eq. (B-65), Eqs. (B-58) through (B-60), and neglecting spatial and orientational gradients in the diffusion tensors, the Smoluchowski equation, Eq. (B-49), can be rearranged to give, for small times,

$$\frac{\partial n}{\partial t} + \epsilon\left(\mathbf{D}_T^0 \cdot \mathbf{F}'^0 + \mathbf{D}_{TR}^0 \cdot \mathbf{T}'^0\right) \cdot \frac{\partial n}{\partial \mathbf{R}} + \epsilon\left(\mathbf{D}_{RT}^0 \cdot \mathbf{F}'^0 + \mathbf{D}_R^0 \cdot \mathbf{T}'^0\right) \cdot \frac{\partial n}{\partial \Phi}$$

$$= \epsilon[\partial/\partial \mathbf{R}\ \partial/\partial \Phi]\,[\mathbf{D}^0] \begin{bmatrix} \partial n/\partial \mathbf{R} \\ \partial n/\partial \Phi \end{bmatrix} \qquad \text{(B-66)}$$

where the 6×6 grand diffusion tensor $\mathbf{D}^0$ is defined by

$$[\mathbf{D}^0] \equiv \begin{bmatrix} \mathbf{D}_T^0 & \mathbf{D}_{TR}^0 \\ \mathbf{D}_{RT}^0 & \mathbf{D}_R^0 \end{bmatrix} \qquad \text{(B-67)}$$

Now, transforming to a set of new variables $(\mathbf{R}', \Phi')$ obtained from the Lagrangian subsidiary system as

$$\mathbf{R}' = \mathbf{R} - \epsilon\left(\mathbf{D}_T^0 \cdot \mathbf{F}'^0 + \mathbf{D}_{TR}^0 \cdot \mathbf{T}'^0\right)t \qquad \text{(B-68)}$$

$$\Phi' = \Phi - \epsilon\left(\mathbf{D}_{RT}^0 \cdot \mathbf{F}'^0 + \mathbf{D}_R^0 \cdot \mathbf{T}'^0\right)t \qquad \text{(B-69)}$$

and letting $\mathbf{q} \equiv (\mathbf{R}' \oplus \Phi')$ be a (1×6) vector, Eq. (B-66) becomes simply

$$\frac{\partial n(q,t)}{\partial t} = \epsilon[\partial/\partial \mathbf{q}]\,[\mathbf{D}^0]\,[\partial n/\partial \mathbf{q}]^\dagger \qquad \text{(B-70)}$$

Note that $[\partial/\partial \mathbf{q}]$ is the (1×6) operator $[\partial/\partial R'_x\ \partial/\partial R'_y\ \partial/\partial R'_z\ \partial/\partial \phi'_x\ \partial/\partial \phi'_y\ \partial/\partial \phi'_z]$ and $[\ ]^\dagger$ denotes the transpose form.

Now, let $\mathbf{S}$ be the 6×6 polar matrix that diagonalizes the grand diffusion matrix—that is,

$$[\hat{\mathbf{D}}^0] = [\mathrm{S}]\,[\mathbf{D}^0]\,[\mathrm{S}]^{-1} \qquad \text{(B-71)}$$

where

$$\hat{D}^0_{ij} = \hat{D}^0_i \delta_{ij} (i, j = 1, 2, \cdots, 6) \tag{B-72}$$

Also, we define

$$[\partial n/\partial \hat{\mathbf{q}}]^\dagger \equiv [\mathbf{S}]\, [\partial n/\partial \mathbf{q}]^\dagger \tag{B-73}$$

Thus, Eq. (B-70) can be rewritten in terms of $\hat{\mathbf{D}}^0$ and $\hat{\mathrm{q}}$ as

$$\frac{\partial n}{\partial t} = \sum_{i=1}^{6} \hat{D}^0_i \frac{\partial^2 n}{\partial \hat{q}_i^2} \tag{B-74}$$

The solution to Eq. (B-74) satisfying the initial condition

$$n = \delta(\hat{\mathbf{q}} - \hat{\mathbf{q}}^0) \tag{B-75}$$

is the multivariate Gaussian function

$$n = \frac{1}{(4\pi t\epsilon)^3 \, \mathrm{Det}[\hat{\mathbf{D}}^0]^{1/2}} \exp\left\{ -\frac{1}{2} [\hat{\mathbf{q}} - \hat{\mathbf{q}}^0][\Sigma]^{-1} [\hat{\mathbf{q}} - \hat{\mathbf{q}}^0]^\dagger \right\} \tag{B-76}$$

where $[\Sigma] = 2t\epsilon[\hat{\mathbf{D}}^0]$. Returning to the variable $\mathbf{q}$ and noting that $\mathrm{Det}[\hat{\mathbf{D}}^0] = \mathrm{Det}[\mathbf{D}^0]$, we obtain

$$n(\mathbf{q}, t) = \frac{1}{(4\pi t\epsilon)^3 \, \mathrm{Det}[\mathbf{D}^0]^{1/2}} \times \exp\left\{ -\frac{1}{4t\epsilon} [\mathbf{q} - \mathbf{q}^0][\mathbf{D}^0]^{-1} [\mathbf{q} - \mathbf{q}^0]^\dagger \right\} \tag{B-77}$$

Equation (B-77) is a multivariate Gaussian function with zero mean and variance-covariance

$$\langle C_i(D^0_{ij}, t) C_j(D^0_{ij}, t) \rangle = 2 D^0_{ij} t\epsilon \tag{B-78}$$

Finally, following Ermak and McCammon (1978), we can use Eqs. (B-77) and (B-78) to put this result in the form of translational and rotational displacement equations as

$$R_i = R^0_i + \epsilon t \left[\sum_j \left(D^0_{ij_T} F'^0_j + D^0_{ij_{TR}} T'^0_j \right) \right] + C_i(D^0_{ij}, t), \quad (1 \le i \le 3\,, \; 1 \le j \le 6) \tag{B-79}$$

$$\phi_i = \phi_i^0 + \epsilon t\left[\sum_j (D^0_{ij_{RT}} F'^0_j + D^0_{ij_R} T'^0_j)\right] + C_i(D^0_{ij}, t), \quad (4 \le i \le 6, \; 1 \le j \le 6) \qquad \text{(B-80)}$$

where $C_i(D^0_{ij}, t)$ is a multivariate, Gaussian random number with zero mean and variance-covariance given by Eq. (B-78). Note that the indices on the components of the diffusion tensors in Eqs. (B-79) and (B-80) run according to Eq. (B-67).

The general BD algorithm in complex systems is straightforward. The body-fixed molecular structure and interaction potential of the B-particle and any nearby surfaces are specified, as well as the bulk solvent density, system temperature and solvent-solvent, solvent-particle molecule, and solvent-surface molecule interaction potentials. For a given initial orientation and position of the B-particle the complete diffusion matrix and equilibrium average force and torque can be determined through equilibrium canonical ensemble molecular dynamics and force autocorrelation analysis. In this step, the grand friction tensor, a numerical array, must be numerically inverted to obtain the grand diffusion tensor. The particle is then moved over a time step t to obtain its new position and orientation. In this step, the forces, torques, and diffusion tensors are assumed approximately constant at their initial values according to Eqs. (B-79) and (B-80). The entire process is then repeated—that is, rehydration and equilibrium MD followed by Brownian dynamics, and the B-particle trajectory is tracked in time.

References

Condiff, D.W., and Dahler, J.S., *J. Chem. Phys.*, **44**, 3988 (1966).

Chandrasekhar, S., *Rev. Mod. Phys.*, **15**, 1 (1943).

Ermak, D.L., and McCammon, J.A., *J. Chem. Phys.*, **69**, 1352 (1978).

Goldstein, H., *Classical Mechanics*, Addison-Wesley, Reading, MA (1980).

Peters, M.H., and Ying, R., *J. Chem. Phys.*, **98**, 6492 (1993).

Peters, M.H., *J. Chem. Phys.*, **110**, 528 (1999).

———., *J. Chem. Phys.*, **112**, 5488 (2000).

———., *Molecular Thermodynamics and Transport Phenomena. Complexity of Scales in Space and Time*, McGraw-Hill, New York (2005).

Wilemski, G. *J. Statist. Phys.*, **14**, 153 (1976).

Index

www.ingramcontent.com/pod-product-compliance
Lightning Source LLC
LaVergne TN
LVHW011519200926
848484LV00001B/78
* 9 7 8 0 0 7 1 4 6 0 7 1 2 *